Computer Algebra and Parallelism

COMPUTATIONAL MATHEMATICS AND APPLICATIONS

Series Editors

J. R. WHITEMAN
Institute of Computational Mathematics, Brunel University, England

and J. H. DAVENPORT
School of Mathematical Sciences, University of Bath, England

E. HINTON and D. R. J. OWEN: Finite Element Programming
M.A. JASWON and G. T. SYMM: Integral Equation Methods in Potential Theory and Elastostatics
J. R. CASH: Stable Recursions; with applications to the numerical solution of stiff systems
H. ENGELS: Numerical Quadrature and Cubature
L. M. DELVES and T. L. FREEMAN: Analysis of Global Expansion Methods: weakly asymptotically diagonal systems
J. E. AKIN: Application and Implementation of Finite Element Methods
J. T. MARTI: Introduction to Sobolev Spaces and Finite Element Solution of Elliptic Boundary Value Problems
H. R. SCHWARZ: Finite Element Methods
J. DELLA DORA and J. FITCH: Computer Algebra and Parallelism

Computer Algebra and Parallelism

J. DELLA DORA
Institut National Polytechnique de Grenoble, France

J. FITCH
School of Mathematical Sciences, University of Bath, UK

ACADEMIC PRESS
Harcourt Brace Jovanovich, Publishers
London San Diego New York Berkeley
Boston Sydney Tokyo Toronto

ACADEMIC PRESS LIMITED
24-28 Oval Road
London NW1

US edition published by
ACADEMIC PRESS INC.
San Diego, CA 92101

British Library Cataloguing in Publication Data is available

ISBN 0-12-209042-X

Printed and Bound in Great Britain by T.J. Press (Padstow) Ltd., Padstow, Cornwall

Preface

Computer algebra, that is the use of computers to perform symbolic mathematics rather than numerical computation, has been an active research area for more than 25 years. For much of that period it was practised by a select group, mainly of applied mathematicians with particular problems to solve, who consumed large quantities of machine resources. It is only in the last five years or so that the availability of large memories on workstations and personal machines has combined with significant mathematical advances to allow computer algebra to reach a much wider audience. With this expansion has come the realisation that there are many problems for which even the most powerful single processor system will not be able to cope adequately. It was the realisation of this limitation, together with a feeling that the time was ripe for an exchange of ideas, that Jean Della Dora and myself were persuaded to call a workshop on computers and parallelism. We were helped greatly in this task by Evelyne Tournier, who made the necessary local arrangements, and assisted in the construction of the scientific program.

So in June of 1988 a group of researchers met at the TIM3 Laboratory of the University of Grenoble to investigate how best to use parallel computing in the field of computer algebra, and also how computer algebra could assist in other uses of parallelism. There were a number of different approaches which we knew were being made in various places. We aimed to bring this diverse group together, to encourage a co-ordinated approach where appropriate, and to establish collaborations. This book contains in a formal scientific format the presentations which were made at the workshop, and as such are complete as a statement of the ideas prevalent at the current time. What the papers do not record are the informal discussions, which are such an important part of a meeting of this kind.

The workshop attracted a broad attendance; established computer algebra specialists and workers in parallel computing, together with research students and young researchers. It was important to the running of the workshop that all those who attended were involved in parallelism in some form, and there was a high degree of participation by all attendees in the discussions. The terms parallelism and computer algebra were interpreted in the broadest sense. This means that in the papers published in this volume can be found descriptions of the way in which the vector capabilities of processors such as the CRAY and Convex can be used in the apparently different style of computation which computer algebra uses, as well as the progress made towards using separate memory, very loosely coupled systems such as could be constructed from a collection of workstations and an Ethernet. Between these extremes there are reported experiments with close-coupled systems, and the possibilities of using pure functional methods. We consider a number of algorithmic developments which are particularly suited for parallel execution, and there is considerable investigation of the possibilities for a parallel Gröbner base method. Other papers consider the necessary software methods needed to realise these algorithms. A slightly different use of computer algebra in parallelism can be found in the use of symbolic computation to write numerical programs for particular parallel architectures. All these papers formed the starting points of discussions.

We were very pleased to notice that a common feature of all the papers was the emphasis on practical methods. The practical work reported was running on actual parallel hardware, and the more theoretical contributions had been tested by implementation. For that reason it was thought that the title of the conference and hence of these proceedings, should have included the word 'practical'. There was an overwhelming emphasis on the solution of real problems and producing usable algebra systems for the future.

As well as the presentations, it had been intended that there should be a short panel session, with a panel taken from the organising committee. In the event it was clearly more appropriate to consider the session as an open discussion, there being a number of views which seemed to differ, and a large proportion of the attendees who would have needed to be members of the panel. The question which was taken as a starting point for the discussion was 'What has been learnt so far in this subject, and are there any general answers?'. It was clear that the topics and techniques which had been advocated had ranged from vectorization to coarse grain. At first sight this would seem to imply a wide and disjointed collection of system issues; in reality the algorithmic concepts are very similar.

This led to a discussion on how we can measure success in a parallel system; there is a need for a set of standard benchmarks which capture real problems. It is not sufficient just to generate artificial problems by recurrence relations, as that is likely to produce results which are invalid in general. Of course it is not just absolute time which matters, but also growth behaviour on families of problems, and the applicability of a particular methodology to a new problem without special tuning. It was also noted that in the early 1970s there was a similar situation in uniprocessor algebra systems, and at that time competition between the various systems had a major beneficial effect. We would welcome test problems for parallel algebra systems.

Another feature of the previous situation was that apart from the first kind of problem, most came from users who presented problems to the experts. At present this field is not sufficiently advanced to attract users, but it was the wish of the meeting that this would change in the near future. We look forward with the hope that before our next meeting there will be a user-led expansion in test problems.

Our discussions identified one technical problem which recurred in a number of projects; that of storage management and garbage collection in a distributed environment. We wish to encourage research into this important area. Of course memory management is an important technique in a wide range of computing, but for the field of general computer algebra it is critical. Without substantial progress in garbage collection in a parallel environment, the requirement for global synchronisation while garbage collecting will form a serious limitation on the development of parallel algebra systems.

The other major need that we identified was for algorithmic development. Computer algebra has made significant progress in recent years because of new algorithmic ideas which have been translated into running code. These algorithms typically have been developed with a sequential model of computing in mind, explicitly or implicitly. In order to make full use of parallel computing in algebra we need new algorithmic ideas. We need algorithms which allow calculations to avoid the need for repeated synchronisation and serialisation. In the proceedings of this workshop can be seen a few initial steps in this direction.

A final remark from our discussion is in order. As well as needing new system and algorithm ideas, we need to be certain when we have seen success! There still needs to be clarification of sustainable speedup, and how general the methods can be. It is all too easy to construct artificial examples which show that with an exponential increase in processors we can achieve polynomial time. We need real achievable speedups on real problems.

It was generally agreed by the attendees that the workshop had been a useful exercise, and one which should be repeated in about two years, when there was hope for progress in all these areas. We are already looking at the timing and location for a second meeting.

The papers in this volume capture the current initial stage on development of *practical* parallelism for computer algebra. As such they describe the base from which we wish to progress. Computer algebra is an interesting and important application area for parallelism. It is more complex than the strict regularity of vector and matrix operations which are so well served by the current generation of supercomputers. Algebraic computations have an irregular structure, but underlying this there is a foundation of mathematical rigour. This makes computer algebra an excellent subject area in which to advance our expertise in parallelism. It is the next step on the path toward completely general parallel computation.

John Fitch
University of Bath

Contents

Contents

Discovering new parallel algorithms. The sieve of Eratosthenes revisited

Michel COSNARD and Jean-Laurent PHILIPPE
LIP - ENSL. 46, Allée d'Italie 69364 Lyon Cedex 07 - France
and
TIM3 - INPG. 46, Avenue Félix Viallet 38031 Grenoble Cedex - France

Abstract

This paper presents three different approaches for generating all prime numbers up to a limit, on parallel computers, based on the sieve of Eratosthenes. We describe Bokhari's algorithm on a *shared memory computer* and discuss the problem of load balancing. We derive a new implementation that improves it.

We then propose a prime number generator and its implementation on a *distributed memory multiprocessor*. We discuss the main problems encountered while parallelizing: problems of communication, load balancing, termination detection and termination of the algorithm. This leads to fully symmetric algorithms for termination detection and termination. We analyze some experimental results obtained on 8 nodes of the hypercube FPS T20, that show a 5-fold speedup. However, the execution time is prohibitive compared with the classical sieve.

We finally show how to efficiently implement the classical sieve of Eratosthenes on a distributed memory multiprocessor. The prime numbers up to 10^8 can be computed in less than 6 minutes, on 16 nodes.

I - Introduction

The practical problem of finding prime numbers still remains almost entire as the power of computers does not follow the needs for such calculations. A possible solution is to use parallel computers. They may either extend the limits or reach the previous limits in a shorter time.

The sieve of Eratosthenes is very simple and therefore highly pedagogical. Dijkstra [Dij 72], Wirth [Wir 73] and Hoare [Hoa 72] use it to explain basic programming concepts. Mairson recalls that the arithmetic complexity of the basic sieve is O(N log log N), but that it can be linearized and even improved to $O(\frac{N}{\log \log N})$ [Mai 77]. Gries and Misra also achieve linear time complexity [GrM 78]. Pritchard shows that the

COMPUTER ALGEBRA AND
PARALLELISM ISBN 0-12-221100-6

algorithms for enumerating prime numbers up to some limit, are sieves (primes are found by computing and eliminating composite numbers) that are linear (each composite is deleted exactly once) [Pri 87].

We recall in the first part of this paper the sieve of Eratosthenes.

We present, in the following sections, prime number generation, first theoretically on a shared memory computer through the sieve of Eratosthenes, and then practically on a distributed memory computer, the FPS T20. On shared memory computers, we discuss the load balancing. On distributed memory computers, many problems have to be solved: communications, load balancing, termination detection and termination itself.

In the last section, we show the implementation of the classical sieve on the FPS T20.

The sieve of Eratosthenes consists of the following steps:

Write down the sequence of integers from 2 to the given limit n.
Take 2 as the first current prime.
Repeat the loop:
Cross out every multiple of the current prime
Update current prime (next uncrossed number)
until next prime is greater than $\sqrt{n}$

To suppress all multiples of a number x, except x itself, we perform only additions: add x to the previous value of the multiple and cross it out.

Many papers have been written on the parallelization of the sieve. [TYN 85] uses it as a benchmark for a new parallel machine: CORAL '83. [HiK 80] presents two parallel versions of the sieve, introducing the problem of synchronization. Bokhari, in [Bok 87], probes the problem deeper and proposes an implementation on a shared memory multiprocessor, the Flex / 32.

II -The sieve of Eratosthenes on shared memory computers

II - 1. Algorithm

Bokhari implements the sieve on the Flex / 32 shared memory multiprocessor with p processors (p=20) consisting of 1 master and p-1 slaves [Bok 87]. This algorithm consists of two sequential steps : writing and sieving.

First step: the master signals each processor in turn to write a sequence of about $\frac{n}{p}$ consecutive numbers in the common memory. Each processor knows (how to compute) the lower and upper bounds of this sequence. When the p-1 slaves have received the order for writing, the master turns into a slave and writes its own sequence. The sequences are balanced, in order to minimize the total time for the writing phase. All the writing processors inform the master when they finish writing.

Second step: the master finds the first uncrossed integer i in the common memory and signals a processor to cross out all multiples of i beginning at 2i. The master finds the next prime i' (the next uncrossed integer) and signals another processor to cross out multiples of i' beginning at 2i', and so on up to $\sqrt{n}$.

The features of this algorithm are: slaves perform concurrent accesses to the memory; global synchronization is done by the master leading to a waste of time; different working speeds could produce temporary dummy prime numbers and extra work.

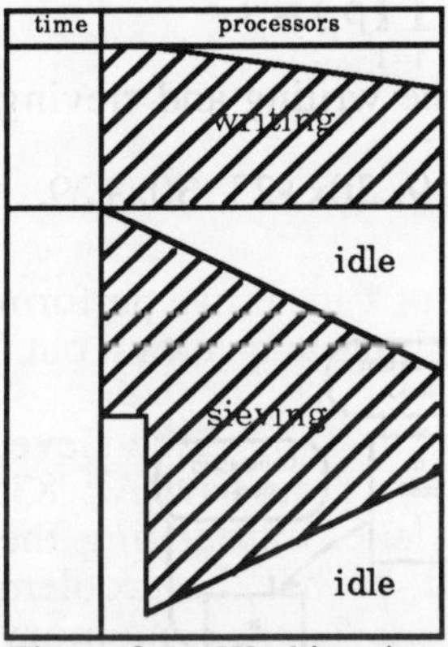

Figure 2-1 : Working time for all processors

If a slave does not cross out the multiples before the master searches for a prime, the master may find an integer not yet crossed out and consider it as a prime. It does not lead to false results but to unnecessary work. The load balancing is bad: there are more multiples of 2 than of 3, 5 or 7... The processor which crosses out multiples of 13, even if it has begun after the one that eliminates multiples of 2, will finish first and may stay idle.

Figure 2-1 shows bad load balancing when there are too many processors compared to the upper limit.

The triangle aspect of the load balancing is not a characteristic of this implementation. But the idleness of many processors and the bad load balancing can be considered as typical.

II - 2. A modification

One of the main limitations of this algorithm comes from the concurrent accesses to the common memory. Assume that this common memory could be divided in several disconnected banks which could be accessed independently by the processors [HwB 84]. In order to eliminate this restriction, each processor writes $\frac{n}{p}$ consecutive numbers in its bank.

The master is the only one to access the whole memory. It *sends the same prime to all the processors* that eliminate the multiples of this number, in their bank.

The basic operation is still addition; there is no concurrent access to the memory during the elimination process; global synchronization is still done by the master; but no temporary dummy prime numbers remain. Load balancing is globally good.

III - Prime number generation on a distributed memory computer

We present a modified version of the sieve of Eratosthenes, implemented on a distributed memory computer, the FPS T20. A description of this computer can be found in [CTV 87].

To eliminate multiples of the first n prime numbers $\rho_1,\ldots,\rho_n$, $(\rho_1=2)$, the general form of the numbers is :

$$\prod_{i=1}^{n}\rho_i \cdot x + y, \quad 0<y<\prod_{i=1}^{n}\rho_i, \quad (y, \prod_{i=1}^{n}\rho_i) = 1$$

To suppress multiples of 2, 3 and 5 before the writing and sieving phases, we generate numbers of the following form :

30x+1, 30x+7, 30x+11, 30x+13, 30x+17, 30x+19, 30x+23, 30x+29.

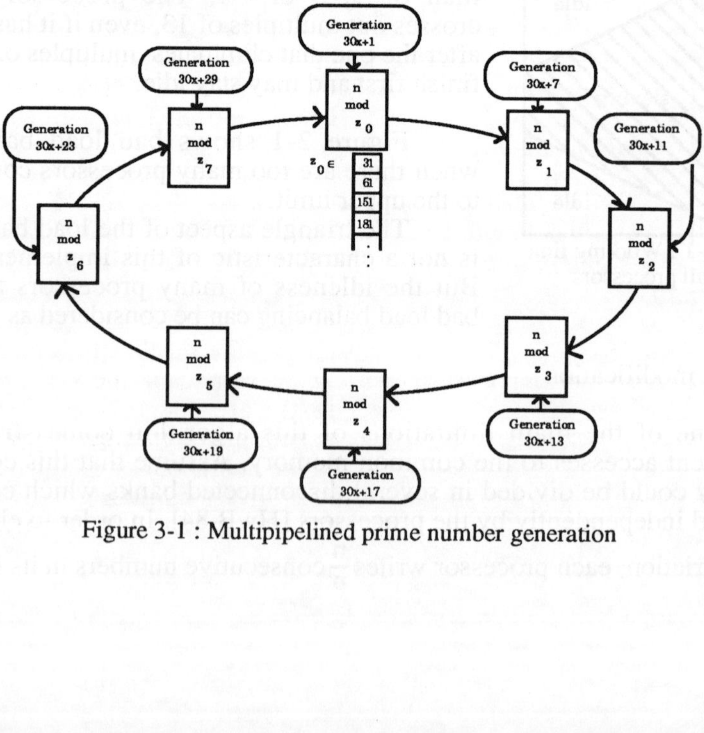

Figure 3-1 : Multipipelined prime number generation

Each processor generates pseudoprime numbers of one of the preceding forms, and tests the local primality of the numbers it receives. The prime numbers are dynamically generated and stored in the local memory by each processor. When an integer has performed a complete cycle on the ring without being eliminated by a processor, it is a prime number and it is stored by the generating processor.

A new number is generated each time the incoming number is not transmitted, either because it is a prime number (stored in the array) or a composite number (eliminated). This technique guarantees that the pipeline is always completely full.

8 processors, one for each y in {1,7,11,13,17,19,23,29}, could be used to generate, test and store the prime numbers.

III - 1. Algorithm

Processor P_i :

```
Receive number n
If n has been generated by Pi then
|      n is prime
|      Store n in the local array
|      Repeat
|      |      Generate m
|      until m locally prime
|      Send m to Pi+1
else
|      If n is prime for Pi
|      |      transmit n to Pi+1
|      else
|      |      Eliminate n
|      |      Repeat
|      |      |      Generate m
|      |      until m locally prime
|      |      Send m to Pi+1
|      endif
endif
```

Figure 3-2 : Algorithm for prime number generation

But the square root of the incoming number may be greater than the largest prime number stored on this processor. To ensure that the algorithm behaves correctly, the incoming number has to be divided by all the prime numbers less than its square root on the processor. This implies the following modification for the program on each processor.

The primality test is done using the prime numbers stored in the local array **AND** the numbers that have been generated but not yet received. However, the greatest generated number may be less than the square root of the incoming number. Hence we introduce a FIFO and use it in the following way. Each time a pseudoprime number is received, the processor checks the rule for the square root. If it is respected, the local primality tests are performed, and the pseudoprime number is either sent to the neighbour or eliminated.

If the rule is not respected, the incoming number is temporarilly queued in the FIFO, while a new pseudoprime number is generated by this processor, replacing the enqueued number. Through this new generation, the value of the numbers that can be tested increases. Each processor can always generate since the following holds :

Let x be the greatest number generated on a processor. The greatest pseudoprime number that can be tested is x^2. The next pseudoprime number that will be generated by this processor will be x+30.

The only case for blocking the generation phase is $x^2 < x + 30$. This is true for $-5<x<6$. These values for x are never reached since $x>30$: the generation phase begins with $30m+y$, $m\geq 1$. □

If the FIFO is empty or if the first number in the FIFO does not satisfy the square root condition, a new pseudoprime number is generated through the usual procedure. On the contrary, if the FIFO is not empty, the first number in the FIFO is dequeued, tested, sent or eliminated.

Remarks: From our experiments, we note that a main difference between this algorithm and the sieve is that non prime numbers are not eliminated in the same order. And, whatever the upper limit is, the program never uses the FIFO (upper limit in the range [100 … 20,000,000]).

III - 2. Results

When the upper limit gets higher, both communication and calculation times increase. But communication time increases slower. However, it remains always greater than calculation time. In the next paragraph, we present a method for decreasing it.

III - 3. A new algorithm

[Saa 85] proposes to modelize the communication time by $t=\beta+\mu q$. [CTV 87] computes the following values of β and μ for the FPS T20: $\beta = 760\mu s$ and $\mu = 1.44\mu s$. From these values, we deduce that it is important to decrease the number of messages. Hence, communicating packets of integers is more efficient than transmitting integers one by one.

Let us group pseudoprime numbers in packets on each processor, and send and receive them. This technique does not introduce new theoretical problems, since generation, primality tests and storage are still performed on the same model and in the same order.

To compute all the primes up to 20,000,000, it takes more than 8,000 s with the first method (one by one transmission) and only 4,000 s with the second method (packets of 1,024 integers).

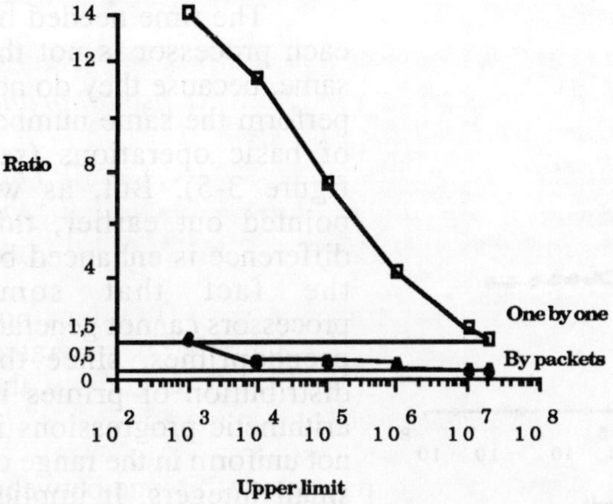

Figure 3-3 : Comparison of $\frac{\text{communication time}}{\text{calculation time}}$ for both methods

Figure 3-3 shows that the ratio $\frac{\text{communication time}}{\text{calculation time}}$ is much smaller with the second algorithm. Moreover, the ratio becomes smaller than 1, which means that the problem becomes compute bound.

This second algorithm is faster than the first one. But processors do not eliminate non prime numbers at the same rate: there are more multiples of 7 than multiples of 19 or 29.

Hence, the processor that eliminates multiples of 7 generates its own pseudoprime numbers (30x + 7) faster than any other processor.

Figure 3-4 shows the number of divisions performed by each pocessor for finding all primes up to 1,000,000.

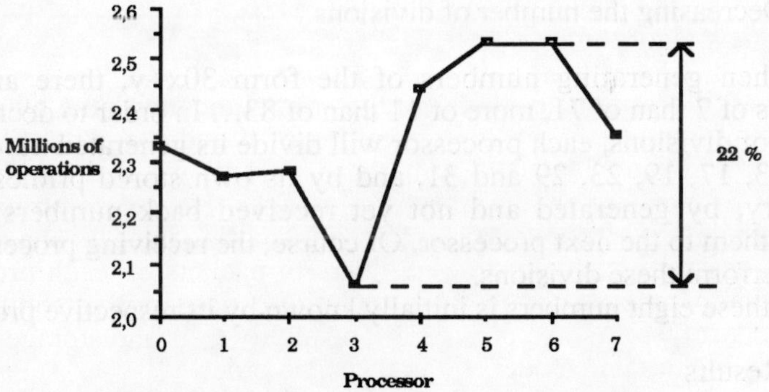

Figure 3-4 : Numbers of operations performed by each processor to find all primes up to 1,000,000.

We can see a difference of about 22% between the processor which performs the greatest number of divisions and the one which performs the least of them.

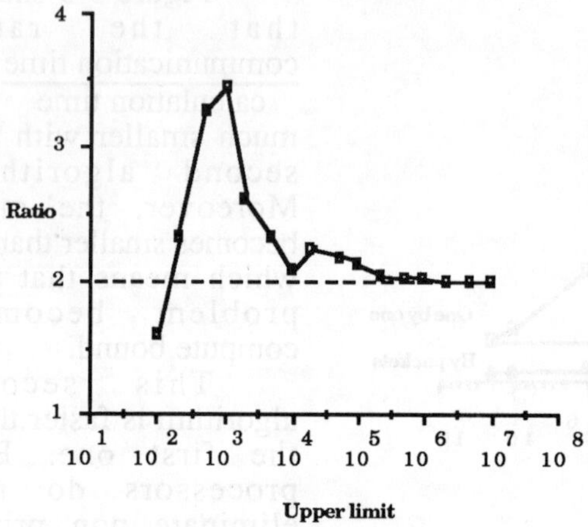

Figure 3-5 : Ratio of the computation time of the slowest processor to the computation time of the fastest processor.

The time needed by each processor is not the same, because they do not perform the same number of basic operations (see figure 3-5). But, as we pointed out earlier, this difference is enhanced by the fact that some processors cannot generate pseudoprimes, since the distribution of primes in arithmetic progressions is not uniform in the range of small integers. It implies that a processor may have no integers to eliminate and no prime numbers to store.

In that case, this processor cannot generate any pseudoprime numbers. We know that this processor will receive a non prime number, in a finite time. On this reception, the generation, test and storage phases will start again for this processor.

III - 4. Decreasing the number of divisions

When generating numbers of the form 30x+y, there are more multiples of 7 than of 71, more of 11 than of 83… In order to decrease the number of divisions, each processor will divide its generated numbers by 7, 11, 13, 17, 19, 23, 29 and 31, and by its own stored primes and, if necessary, by generated and not yet received back numbers, before sending them to the next processor. Of course, the receiving processor will not re-perform these divisions.
Each of these eight numbers is initially known by its respective processor.

III - 5. Results

The number of operations decreases of approximately 30% (see figure 3-6), which is a good improvement for such a minor modification of the algorithm and such a small redundancy (8 integers). The third method is almost twice as fast than the previous one (see figure 3-7).

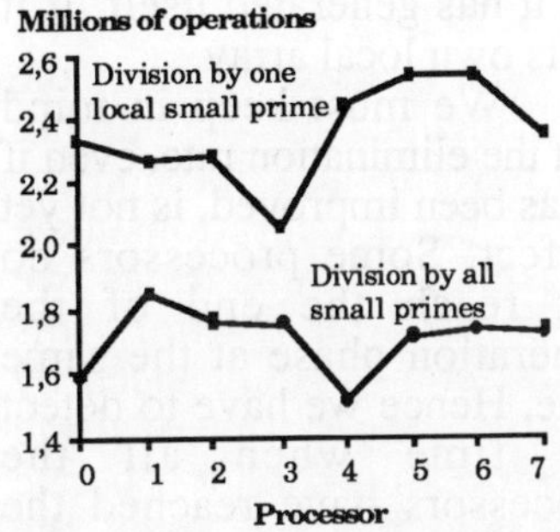

Figure 3-6 : Gain on the number of operations to enumerate all primes up to 1,000,000.

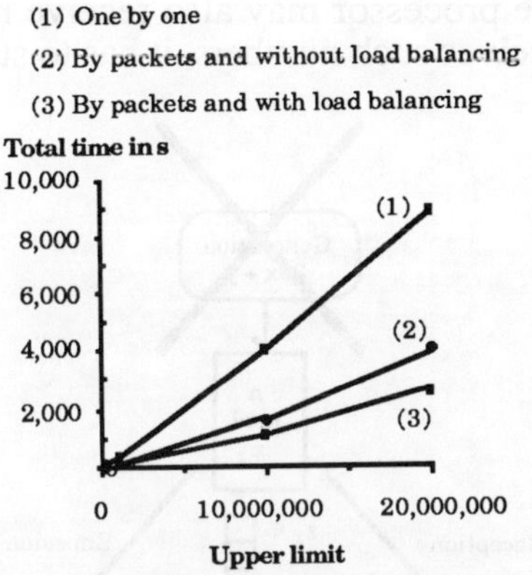

Figure 3-7 : Comparison of total execution times using the three methods.

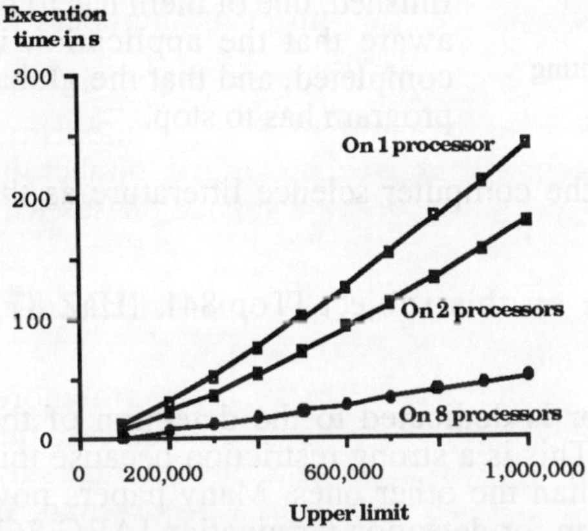

Figure 3-8 : Execution times of the best algorithm on 1, 2 and 8 processors

Let us compare the execution times of this third algorithm on one processor (no communication, only computation), on a ring of two processors (6x+1 and 6x+5) and on a ring of eight processors (30x+y).

We can see that, in the range 100,000 - 1,000,000, executing the algorithm on an eight and even on a two processors ring saves time, compared with an execution on a single processor. The maximum speedup is greater than 5 for 8 processors with an upper limit of 1,000,000 (see figure 3-8).

IV - Distributed termination

A processor stops when the next number that it could generate is higher than the upper limit n every processor knows. When a processor stops, it does not generate pseudoprime numbers anymore. But it still tests the primality of incoming numbers. If the incoming number is prime for this processor, it is sent to the next one. If it is not prime, it is eliminated.

The processor may also receive numbers that it has generated itself. If it receives such numbers, it has to store them in its own local array.

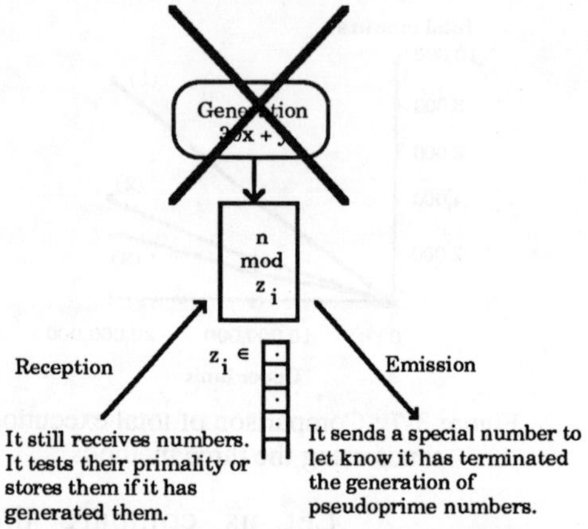

Figure 4-1 : A processor stops generating

We must keep in mind that the elimination rate, even if it has been improved, is not yet perfect. Some processors do not reach the end of the generation phase at the same time. Hence we have to detect the time when all the processors have reached the upper limit. The processors that do not generate pseudoprime numbers anymore must let the other processors know. Moreover, when all of them have finished, one of them has to be aware that the application is completed, and that the global program has to stop.

This problem is known in the computer science litterature as the termination problem.

Much work has been done on this subject [Top 84], [HaZ 87], [Ray 85], [SaR 86].

In some papers, a processor is dedicated to the detection of the termination [Top 84], [DFG 83]. This is a strong restriction because this processor has much more work than the other ones. Many papers now present a fully symmetric algorithm for detection termination [ARG 86], [Ran 83], [HaZ 87], [Bla 88]. In this class of algorithms, any processor has the same right for detecting termination.

A second restriction is to know if the messages must be kept in the chronological order. This introduces the need for a FIFO. Dijkstra presents an algorithm that does not assume the FIFO property for the communications [DiS 80]. But this algorithm suffers from another severe property: the number of messages for detecting the termination is the number of messages that the computation has needed all over the execution. Moreover, it slows down the execution of the main algorithm. Another class is the termination detection algorithms with markers [Fra 80], [Mis 83]. Most of these papers include a proof of the presented algorithms. But other papers are dedicated to the proof of distributed termination algorithms [Apt 84], [Ver 87].

IV - 1. Algorithm

We now present a fully symmetric algorithm which includes the termination algorithm itself. We shall prove its correctness.

The messages perform the following functions: termination detection and termination of the algorithm. p is the number of processors on the ring. There are two types of messages:

- pseudoprime numbers (the processors test their primality): *strictly positive values.*

- an ending number which contains a counter. *The value of this special number is in the range [-2p ... -1].* This number indicates to every processor that one of them has stopped generating pseudoprime numbers. The processor, when stopping, sends this counter with the value -1. Firstly, it may be replaced by a pseudoprime number generated on the next first active processor. If the receiving processor is passive, it adds -1 to this counter and sends it to its successor. Secondly, when this ending number has performed one complete cycle on the ring (the value of the counter is -p), then no processor is still active. The receiving processor detects the termination and begins the proper termination phase. Thirdly, this counter empties the pipelines and terminates all the communications: processors know that they must begin this phase through the value of the counter which is less than -p. In this termination phase, processors go on adding -1 to the counter. It keeps the coherence with the detection phase.

The counter is not present on each processor. But it is contained in the message that tries to detect global termination.

More than one processor may begin the detection phase at the same time, detect the termination and run the termination algorithm. It takes p communications since the counter value must reach -p so that a processor detects termination. In the same manner, it takes p communications to complete.

The corresponding automaton is presented below.

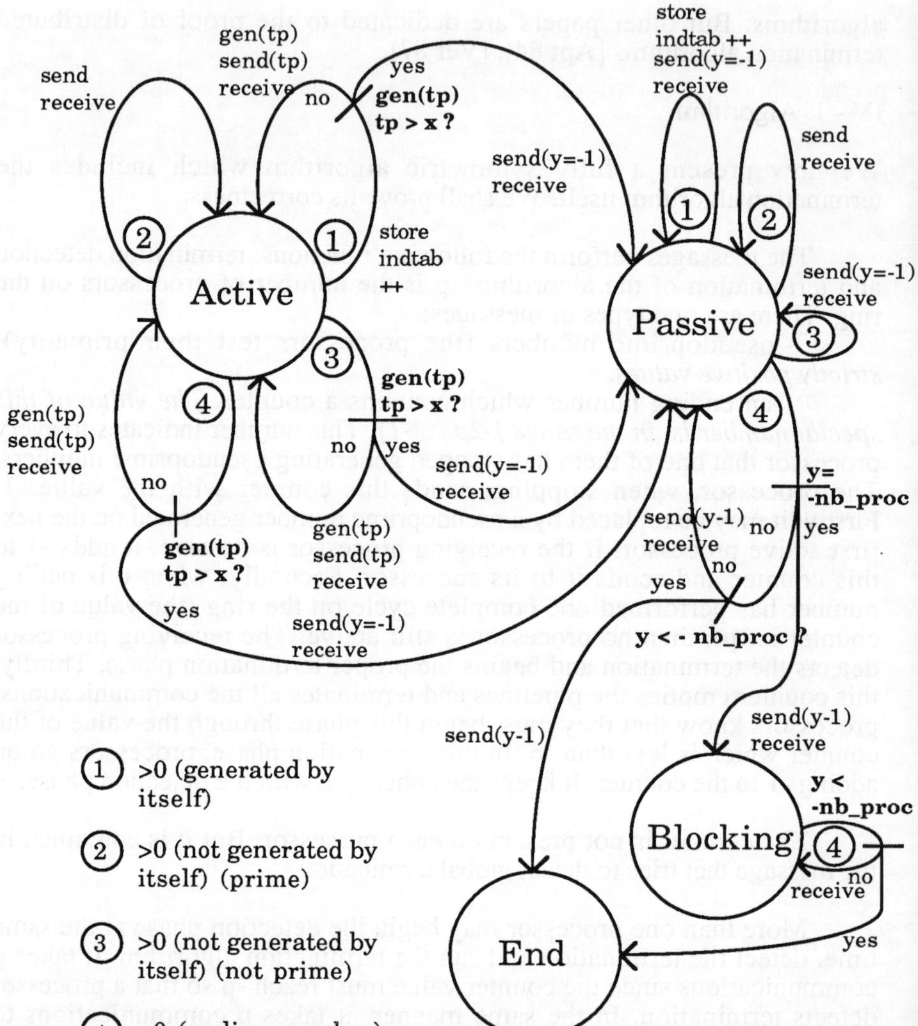

Figure 4-2: Automaton of the detection termination and termination algorithm

IV - 2. Proof of correctness of this distributed termination detection algorithm

In our algorithm, the concepts for active and passive states are the following ones. When a processor becomes passive, it cannot go back to the active state. However, the passive state for a processor does not mean that it does not work anymore. It only means that it will no longer generate pseudoprimes. But it still tests the primality of the incoming numbers.

Two assertions have to be proven (following a skeleton proof given by [HaZ 87]).

1) *The termination of all the running processes will not go undetected.*

Proof (in a non formal style) :

When a process (there is a bijection between processes and processors) reaches the upper limit n, it sends a wave. This wave is a negative counter initialized to -1. It is sent to the next processor on the ring. If the receiving processor is still active, it deletes this wave. If it is already passive, it adds -1 to the counter and sends it to the next processor on the ring. During the execution of the program, all the processors initiate this termination detection. But the wave could be stopped by a still active processor.

At a given time, there remains only one active processor. It will reach the upper limit and then send -1. This counter will perform a complete cycle on the ring and will be regularly decremented by each processor. The counter comes back to its sender equal to -p.

Distributed termination is therefore globally detected.

□

2) *The global termination is not detected if it has not occurred.*

Proof (in a non formal style) :

The global termination is detected only if a processor receives a counter with a value less than -p. Let us prove that such a counter value cannot occur while all processors have not terminated. When a processor locally terminates, it sends a counter equal to -1. This counter is sent to the next processor on the ring. If this processor is active, the counter is eliminated. If the processor is not active anymore, it decrements the counter and sends it. It is not possible for a processor to receive a counter with a value less than or equal to -p if a processor is still active, because this processor would have deleted the counter.

So the global termination is not detected if it has not occurred.

□

IV - 3. Distributed termination conclusion

The features of our algorithm are the following ones: one ring cycle is necessary to detect the distributed termination. One cycle is necessary to terminate (empty the pipelines and close the communications) once the termination has been detected. The algorithm is fully symmetric since all the processors begin the termination detection algorithm. Any processor may detect the termination. When one of them has detected it, it begins the termination algorithm. More than one processor can begin the termination algorithm independently. This algorithm is very simple, since it does not introduce a superimposed topology and a special control algorithm.

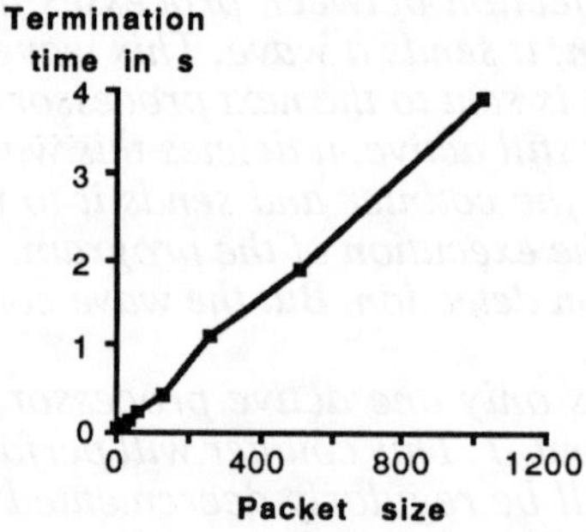

Figure 4-3 : Termination time in function of packet size

Figure 4-3 shows the termination time for a processor on the ring of 8 processors when enumerating all primes up to 20,000,000, in function of the size of the packets of pseudoprime numbers that are transmitted.

V - The classical sieve of Eratosthenes on the FPS T20

Our algorithm is derived from the one presented in section 2-2. We assume that the memory is distributed among the processors. The lower and upper bounds of the sequence depend, as in Bokhari's algorithm, on the identity of the processors. On the FPS T20, each processor is identified by a number ranging from 0 to 15. Let p_i be the identity of processor i and n be the upper limit for enumerating prime numbers. p_i writes integers from $(1+p_i \frac{n}{16})$ up to $(p_i+1) \frac{n}{16}$ in its local memory.

No master can access the entire global memory. A special processor will find the next prime number and send it to all processors. Then each processor will suppress multiples of this number. Of course, the special processor will stop looking for the next prime when it reaches $\sqrt{n}$.

V - 1. Implementation

We configurate the 16 processors of the hypercube FPS T20 as a ring and use the one-to-all communication procedure to send the next prime, found by processor 0, to all the other processors.

The numbers are logically present in an array of booleans on each processor. The numbers that we sieve are the indices of these arrays.

Each element of the arrays is initialized to true (each number is initially a prime number). Then each time an integer is crossed out, the value corresponding to its index in the array is turned to false. Composite numbers are turned many times to false. This implies redundant work. The primes are the integers corresponding to the indices for which the boolean values are still true at the end of the sieving phase.

On each processor, we use an array of 200,000 integers (800 Kb) containing 32 x 200,000 = 6,400,000 bits. On 16 processors, the greatest upper limit n is 102,400,000.

V - 2. Results

Upper limit	Execution time
1,024	0.0148 s
10,240	0.0548 s
102,400	0.355 s
1,024,000	3.26 s
2,048,000	6.54 s
4,096,000	13.2 s
10,240,000	33.5 s
20,480,000	67.8 s
40,960,000	137. s
102,400,000	350. s

Figure 5-1 : Execution time of the sieve on 16 processors

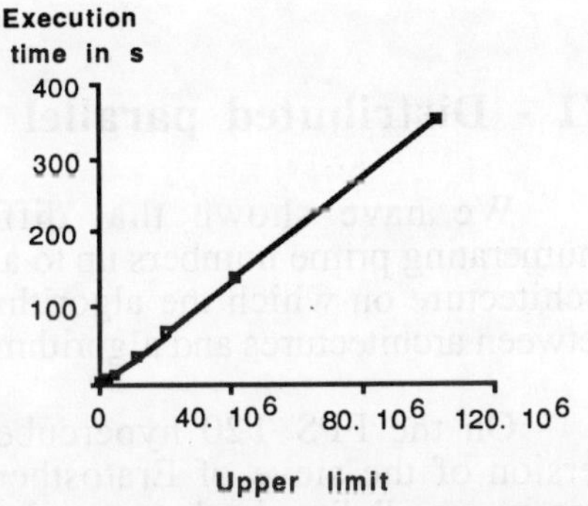

Figure 5-2 : Execution time of the sieve.

The execution time is close to be linear (see figure 5-2).

When comparing, we can see that the classical sieve is much faster than the algorithm studied in section 3 for distributed memory architectures (see fig. 5-3).

Upper limit	Execution time Prime number generator	Execution time Sieve	Ratio
1,000	0.0316 s	0.0147 s	2.1
10,000	0.198 s	0.054 s	3.6
100,000	1.42 s	0.352 s	4.0
1,000,000	52.5 s	3.24s	15
10,000,000	1580. s	33.4 s	47
20,000,000	4031. s	67.8 s	59

Figure 5-3 : Comparison of the sieve of Eratosthenes and the best version (packets and good load balancing) of the prime number generator

V - 3. Conclusion

The sieve of Eratosthenes is very efficient in its classical version and when correctly using the memory (one bit per boolean). It then extends the limits for computation. These limits were, in the previous versions, bounded by two factors : memory size and execution time. Now execution time has drastically decreased. And the limits depend now only on memory size.

VI - Distributed parallel algorithms conclusion

We have shown that different approaches are possible for enumerating prime numbers up to a limit. These approaches depend on the architecture on which the algorithms will be implemented: interrelation between architectures and algorithms is very strong.

On the FPS T20 hypercube, we have implemented a modified version of the sieve of Eratosthenes. We have faced many problems: massive parallelism leads us to substantially modify sequential algorithms. Communications, load balancing, termination detection and termination itself are problems that must be taken into account. We have tried to show the methodologies we have used and the logical reasoning we have followed to derive the final algorithms.

Though we have proven the correctness of the termination detection algorithm, proving parallel algorithms is very difficult.

But parallel algorithms can be very efficient !

References

[Apt 84] APT, K.R. (1984). "Correctness Proofs of Distributed Termination Algorithms", Tech. Rep. 84-51, LITP, University Paris VI.

[ARG 86] ARORA, R.K., RANA, S.P., GUPTA, M.N. (1986). "Distributed Termination Detection Algorithm for Distributed Computations", I.P.L., 22, 311-314.

[Bla 88] BLANC, P. (1988). "Distributed termination detection when messages arrive out of sequence", Conf. on Par. Proc., Pisa, Italy, 25-27 April 1988.

[Bok 87] BOKHARI, S.H. (1987) "Multiprocessing the Sieve of Eratosthenes" IEEE Computer, 20, 4, 50-58.

[CTV 87] COSNARD, M., TOURANCHEAU, B., VILLARD, G. (1987). "Présentation de l'Hypercube T20 de FPS", Bigre + Globule, 56, 12-17.

[DFG 83] DIJKSTRA, E.W., FEIJEN, W.H.J., VAN GASTEREN A.J.M. (1983). "Derivation of a Termination Detection Algorithm for Distributed Computations", I.P.L., 16, 5, 217-219.

[Dij 72] DIJKSTRA, E. W. (1972). "Notes on Structured Programming", In "Structured Programming", 1-82, Academic Press, New York.

[DiS 80] DIJKSTRA, E. W., SCHOLTEN, C.S. (1980). "Termination Detection for Diffusing Computations", I.P.L., 11, 1, 1 4.

[EdH 85] EDWARDS, J.R., HARTWIG, G. (1985). "Benchmarking the Clones", Byte, 10, 11, 195-201.

[Fra 80] FRANCEZ, N. (1980). "Distributed Termination" ACM Trans. on Progr. Lang. and Syst., 2, 1, 42-55.

[GrM 78] GRIES, D., MISRA, J. (1978). "A Linear Sieve algorithm for Finding Prime Numbers", Com. ACM, 21, 12, 999-1003.

[HaZ 87] HAZARI, C., ZEDAN, H. (1987). "A Distributed Algorithm for Distributed Termination", I.P.L., 24, 293-297.

[HiK 80] HIKITA, T., KAWAI, S. (1980). "Parallel Sieve Methods for Generating Prime Numbers", 257-262, IFIP, S.H. Lavington Editor, North-Holland Publishing Company.

[Hoa 72] HOARE, C.A.R. (1972). "Notes on Data Structuring", In "Structured Programming", 83-174, Academic Press, New York.

[Hoa 78] HOARE, C.A.R. (1978). "Communicating Sequential Processes" Com. ACM, 21, 8, 666-677.

[HwB 84] HWANG, K., BRIGGS, F.A. (1984). "Computer Architecture and Parallel Processing", 846 p., McGraw Hill Book Company.

[Mai 77] MAIRSON, H. G. (1977). "Some New Upper Bounds on the Generation of Prime Numbers", ACM, 20, 9, 664-669.

[Mis 83] MISRA, J. (1983). "Detecting Termination of Distributed Computations Using Markers", ACM SIGACT-SIGOPS Symp. on Princ. of Distr. Comp., Montreal, Canada, August 17-19.

[Pat 82] PATTERSON, D.A. (1982). "A Performance evaluation of the Intel 80286", Comp. Arch. News (ACM SIGArch Newsletter), 10, 5, 16-18.

[Pri 87] PRITCHARD, P. (1987). "Linear Prime-Number Sieves : a Family Tree", S. of Comp. Progr., 9, 17-35, North Holland.

[Ran 83] RANA, S.P. (1983). "A Distributed Solution of the Distributed Termination Problem", I.P.L., 17, 43-46.

[Ray 85] RAYNAL, M. (1985). "Algorithmes Distribués et Protocoles", 140 p., Editions Eyrolles, Paris.

[SaR 86] SAIKKONEN, H., RONN, S. (1986). "Distributed Termination on a Ring", Bit, 26, 188-194.

[Saa 85] SAAD, Y. (1985). "Communication Complexity of the Gaussian Elimination Algorithm on Multiprocessors", R.R. 348, Comp. Sc. Dep., Yale University.

[Sch 86] SCHROEDER, M.R. (1986). "Number Theory in Science and Communication", Spr. Series in Inf. Sc., 374 p., 116-117, Springer Verlag, Second Edition.

[Top 84] TOPOR, R.W. (1984). "Termination Detection for Distributed Computations", I.P.L., 18, 1, 33-36.

[TYN 83] TAKAHASHI, Y., YAMANE, Y., NISHIYAMA, K., YOSHITANI, F., INOUE, K. (1985). "Efficiency of Parallel Computation on the Binary-Tree Machine CORAL '83", Journ. of I. P., 8, 4, 288-299.

[Ver 87] VERJUS, J.P. (1987). "On the Proof of a Distributed Algorithm", I.P.L., 25, 145-147.

[Wir 73] WIRTH, N. (1973). "Systematic Programming : an Introduction", Prentice-Hall, Englewood Cliffs, New Jersey.

Compiling For Parallelism

John Fitch
School of Mathematical Sciences
University of Bath
Bath BA2 7AY, England

1. INTRODUCTION

There are many ways in which it is possible to start the quest for parallelism in an algebra system. The work described here is based on a small number of clear criteria which have shaped the whole project. The first section of the paper presents these and gives the justification for them. The details of the compilation for parallelism form the subject of the second section. There are a number of basic enabling technologies on which our work is based, and these are explained here. The third part of the paper describes briefly the multiprocessor LISP system we have constructed. The paper finishes with some assessment of how far we have got, and some speculation on the future of the project.

Part of the work described here was developed in Bath, a contribution to the cost being made by the Science and Engineering Research Council, and part is closely related to the Concurrent Processing for Advanced Simulation (CPAS) project at the RAND Corporation. I wish to record the part played by Jed Marti of RAND in writing the first data flow analyser, and in helping me in refining my ideas on its development. The multiprocesssor LISP is largely the work of Julian Padget.

COMPUTER ALGEBRA AND
PARALLELISM ISBN 0-12-221100-6

2. DESIGN CRITERIA

The desire for some kind of intelligent compilation came from the algebraic manipulation work at the University of Utah in the 1970s and the work we did in investigating how one could use parallelism. This lead to an important guiding principle of the work; there was a significant body of code in the REDUCE system which incorporated a number of algorithms, mathematical theorems and pragmatic ideas. The algorithmic ideas were being developed, but the systems being used were not getting faster to the extend we needed; a processing power gap was being opened. This led directly to the desire for a parallel LISP system. But the amount of code already written was a considerable investment, and so we could not contemplate the rewriting of the whole of REDUCE. This is our first design axiom: **the parallel system must accept code written before with no (or very little) change.**

The second axiom is justified by the unlikelihood of our being able to raise large sums of research money. This means that we seek a parallel scheme that can be constructed from off-the-shelf components rather than specialised machines. This gives us **use only standard hardware**.

While creating new hardware is beyond the resources of the project, we are practised in system software, and in writing LISP in particular. Accordingly all the code related to this project has been either based on Cambridge LISP [Fitch and Norman, 1977] or Portable Standard LISP [Griss, Bensen and Maguire, 1982], both of which are the base on which we run REDUCE.

These two rules have formed a major plank of our platform. We are striving to construct a non-shared memory, loosely coupled software architecture for a modest number of processors to cooperate on a single algebraic computation. This architecture may benefit from specialised hardware, but we cannot rely on any such support.

Another component of our search for a viable parallel execution model for algebra was an unsubstantiated feeling that what was needed was not lazy evaluation but rather some form of haste. The earliest essays at this form are in Marti [1980a, 1980b] and the Bath Banque [Bowyer, Willis and Woodwark, 1981][Marti and Fitch, 1983].

3. DETECTION OF CONCURRENCY

If we are to achieve our goal of parallel execution of pre-existing programs then we need a suite of analysis tools which investigate the source program statically, looking for the available coarse-grain concurrency, and in particular parallelism at the level of a function. This main section of the paper considers the tools, their design and efficiency. The two planks are data flow analysis and static runtime estimation.

3.1. Flow Analysis

If we were using pure LISP or any other functional language suxh as ML then we could be certain that there could be no interference between the evaluation of two expressions. In fact the style of real LISP used in the writing of REDUCE shows a large component of this functional origin, but on many occasions use is made of updating, and global assignment is a major mechanism for control. The basic idea for detecting potential parallelism is the data flow analyser, which has its roots in compiler construction. This takes functions, or other fragments of code, and constructs a flow graph from it, using a straightforward method of code walking. This is then combined with semantic information about the basic LISP functions and rules for the various methods of composition to give a semantic description of the side effects of the function. In effect this description is an annotated closure of the function which identifies the non-local effects of the function. We will see latter how this closure is used.

The special features that the annotations identify are

- *read only*

  those non-locals whose value is used but not modified

- *read/write*

  non-locals whose values are used and modified, and so the code requires exclusive use

- *write only*

  values modified by the function, before their value is used if at all.

To these three sets is added a flag to indicate "hard" functions, by which we mean those which are to hard to analyse at compile time; example of hard functions include some instances of the functions **set** and **nconc**. As the annotated closures have four fields, they are known as *Rquads*.

With these semantic forms we introduce a number of operators. The simplest conceptually of these is *BNEAR* which is a boolean function determining if there is an overlap between two sets, whether they be read, read/write or write. The definition of this function is fairly straightforward. From this predicate we can define set-like operations such as inclusion, difference, intersection and union. These give basic building blocks for the binary constructors of rquads from other rquads:

-> gives the side effect set of evaluating A and then B;

<> gives the side effect set of evaluating A before B, or of evaluating B before A.

A definition of -> slightly simplified is given in figure 1.

These operators are used by the fundamental analyser to combine function calls, and to evaluate arguments retaining the left to right order which is used in REDUCE. If the program were presented in an order where each function is defined before it is used, then these operators would be sufficient to

```
(de -> (A B)
  (cond ((equal A B) A)
        ((or (equal A Rhard) (equal B Rhard)) Rhard)
        ((equal A R0) B)
        ((equal B R0) A)
        (t (make-rquad
              r    (\/ (-- (rquad-r A)
                           (\/ (rquad-w B) (rquad-rw B)))
                       (-- (rquad-r B)
                           (\/ (rquad-w A) (rquad-rw A))))
              w    (\/ (rquad-w A)
                       (-- (rquad-w B)
                           (\/ (rquad-r A) (rquad-rw A))))
              rw   (\/ (rquad-rw A)
                       (\/ (-- (rquad-rw B) (rquad-w A))
                           (/\ (rquad-r A) (rquad-w B))))
              hrd    false))))
```

Figure 1: Simplified Definition of Side Effect Set Operator ->

determine the all important side effect sets. This order would be too great a restriction on the programmer, especially as we wish the bulk of the code to be unchanged. In order to allow the program to be presented in any order an evaluation strategy for rquads is used which mixes explicit rquads and algebraic forms, followed by a "back substitution" pass to complete the analysis. Some additional technology is needed to deal with "spaghetti" code, and the semantic analysis of procedural segments with jumps as found in the PROG feature poses additional complication. This detail is not presented here but is left for Marti and Fitch[*in preparation*].

The provision of the hard flag permits a particularly easy analysis, in which every function is marked as hard. This is the ultimate state, and the data flow analyser attempts to improve on this, but is cases of doubt it always takes the safe interpretation. The danger is that all the results tend to this form. In practice this happens rarely in the code we have tested so far. A more significant problem is the treatment of data other than simple variables. The first main class of these in LISP is the property list which is extensively used in REDUCE. One can view this as a location indexed by two names, the identifier and the indicator. We model this simply by extending the elements of a set from simple names, to include name pairs. An added complication is that we rarely know both the identifier and the indicator, but for the standpoint of side effect computation, even if the identifier is not known, different indicators ensure that there is no clash. The introduction of an *unknown* symbol is necessary. The basic components of the data flow analyser are quite small; we have analysed some non-trivial programs on a 64Kb Z80. Major sections of REDUCE are analysed on an Orion 1.

The current implementation of REDUCE is written in a fairly simple style of LISP, but modern LISPs contain record structures, which need to be acceptable to the analyser. Indeed the data flow analyser itself uses the **defstruct** construction. Extending in this way is quite easy, and a naive treatment of vectors is also possible. The basic system has been extended beyond what is required for algebra; part of the funding is for object oriented programming, and an extension to handle an object system with simple inheritance has been constructed, details of which are in Fitch [1986].

In order for the data flow analyser to start we need semantic information for the basic functions of LISP, which indicate any side effects they entail. A number of functions, such as **car** and **member** are clearly side-effect free, and can be coded with the null rquad (which we call R0). Others are hard, but many can be assigned particular behaviour. Printing functions can be encoded as writing to variables which give the position on the page. To date we have written semantic functions for both PSL and Cambridge LISP.

An example of the use of the data flow tool is given in figure 2, which shows a simple collection of functions, and the results of the analysis. The star symbol refers to a value cell, and the question mark is an unknown value. The number in brackets is the cost analysis, which is described in the next section. The analysis shows that the functions **bar** and **baz** could be performed in parallel if the cost analysis shows it is worthwhile.

```
(fluid '(xx yy zz))

(de foo (xx)
   (list yy (bar xx) (baz xx)))

(de bar (p)
   (setq zz (get p 'tag))
   (put zz 'ind p)
   42)

(de baz (q)
   (prog (ans)
         (setq ans yy)
         (get yy 'other)
         (setq yy q)
         (return ans)))

-- Analysis of example begins --

baz(3) :: [read:{?:other} write:{} read/write:{yy:*}]

bar(6) :: [read:{?:tag} write:{zz:*, ?:ind} read/write:{}]

foo(9) :: [read:{?:tag, ?:other} write:{zz:*, ?:ind} read/write:{yy:*}]

-- Analysis ends --
```

Figure 2: Example Input and Output for Data Flow Analysis

3.2. Cost Analysis

The data flow mechanism described in the last section can be used to show where the concurrent execution of two fragments is semantically equivalent to their evaluation sequentially. This is not sufficient. In a practical system, with real communication delays, we must not create a new process, or transmit a task for a calculation which would complete faster than the cost of transmission of the question and the return of the answer. In order to decide when a distribution of work could be beneficial we have constructed a heuristic cost analysis as part of the static analysis.

If we can estimate statically the dynamic behaviour of a function, and in particular the run time, then we are in a position to make an intelligent decision. The detailed decision depends on the relative costs of transmission and the amount of processing time needed for a given portion of the computation. This is, in turn, related to the granularity of parallelism in the computation: if the program is broken into very many small tasks, each of whose running time is very small, then it is possible that the amount of processing required to transfer the code and its closure from one processor to another and back again will exceed the time needed to execute the program sequentially on a single processor.

It is clear that a *calculation* of the dynamics of a program includes the halting problem as a subset, and so is undecidable. This pessimistic answer does not take account of two points; firstly we are not analysing random programs but rather programs written by humans to perform real tasks, and secondly we only need an estimate. The quality of our estimate will have a significant effect on the performance of the eventual system, but the effect is only on performance.

At the same time as the data flow analyser is constructing the control graph it also maintains information on the costs of evaluating each arc. This is built from adding up the costs of the basic building blocks. In order to do this we need know the cost of the elementary LISP functions, which we consider in the next paragraph. The difficulties in the cost analysis are the behaviour of loops, recursion and conditionals. With conditionals, we can use either the maximum or average of the various routes, and we have experimented with both of these. A better method would be to use knowledge of run time to us a weighted average, but we have not done this yet. The case of loops is considerably more complex. In simple *for* loops it may be possible to know how many times the body is obeyed, but most loops, and the implicit loops in spaghetti code, are or their nature unknown. We are using a chance remark from one of the designers of MU5, that the average number of times a loop is obeyed is eight, to provide a guess for this unknowable. This value does seem to agree fairly well with our measurements.

We have used two approaches to providing data for the static estimation. The first and simplest was to make informed guesses as to the relative times for the basic functions; a more complex method has been to make measurements of actual run times in a test bed. We have done this for one particular PSL. The agreement between the predicted run time and the measured time was remarkably good, especially with the first method, being always within a factor of two, and often very much closer. This subject is still a subject of new research.

3.3. Local and Global Side Effects

In experimenting with the data flow system we came across a style of code which was defeating the analysis. This was because the data flow analysis will treat as *hard* any function whose behaviour cannot be determined. The means that if any function calls a hard function, then this will in its turn be hard. This effect can lead to most functions being hard, and so lead to global synchronisation always being requested. While it is pleasing that in general this does not happen, the style of building a list backwards, and then reversing the list destructively on exit (as for example in figure 3) gives an unnecessary synchronisation; the operation is in fact safe.

This effect was sufficiently severe in some of our programs that we were forced to consider extending the analysis to covering these cases. In fact we know in this example that the destructive reverse **nreverse** is only on new structure, created within the function. By tracing which variables hold new unshared structure we can replace the **nreverse** with a function **pure-**

```
(de add-one-to-list (l)
    (let ((ans nil))
         (foreach x in l do (setq ans (cons (add1 x) ans)))
         (nreverse ans)))
```

Figure 3: Simple Function Showing Local Destructive Operation

nreverse which has the same actions, but has an R0 semantic expression. This action is similar to abstract interpretation which is used in strictness analysis for lazy evaluation. In fact the system described here was developed independently of abstract interpretation, and it was only recently that the similarity was noticed. We have detected and dealt with calls to a number of these safe destructive operations.

This approach to understanding the program is only at the beginning. With deeper analysis it may be possible to detect other independent actions. We are actively pursuing this approach.

3.4. A New Compiler

The next stage which we have only just begun is to make use of this analysis in a compiler. We expect to use a simple FORK-JOIN style of process synchronisation, although we are also experimenting with the Time-Warp methodology. What we have done so far is to lay down the frame work on which we can build a compiler.

The compilation process we are currently constructing relies on the semantic information determined by the data flow analysis. In a function call if all the arguments involve Rquads which have a null intersection then all the arguments may be evaluated in parallel if the static analysis indicates that this is desirable. If the intersection is not null then the side effect sets impose a partial ordering on the arguments if we are to preserve the left-to-right evaluation semantics usual in LISP. This applies not only to conventional functions and user defined functions, but also to explicitly sequential functions such as PROGN or even PROG with a little additional care. Extension of this idea to the mapping functions is clearly possible. The compiler uses this information to determine the rendezvous points within each function. It may be possible to incorporate the *future* ideas from MultiLISP [Halstead, 1985].

4. MULTIPROCESSOR LISP

The previous section has described an approach to the detection of parallelism in existing LISP programs in order that we can use it in compilation. We still need a LISP capable of exploiting multiple processes.

To have each processor occupied by a single process is clearly an inefficient use of resources, requiring more processors than we can afford, and so we want each LISP processor to be multi-tasking. For practical reasons it is reasonable to have the management of all the machine's resources under the control of the LISP system. That means the LISP system is responsible for handling all communications, process creation, destruction and scheduling. The alternative would be to extend the operating system with a remote LISP evaluation daemon which would fire up a new LISP process for each evaluation request received; this would be more profligate with resources (a workable LISP image probably has to be more than 2Mb, with a significant working set and besides the duplication of all the basic LISP functions). Also as we do not have access to the source of the operating system, which we would need to get processes scheduling right, and we are all LISP speakers, the easy option is to build a multi-tasking LISP. This section gives a short description of some of the features of the system being constructed.

4.1. Multiple Control

If a single LISP is to evaluate a number of processes, then we need to be able to retain the control environment. In the LISP literature there are a number of ways this can be done. The approach we have adopted is to use stack-groups as in the LISP machine [Weinreb and Moon, 1984]. This is well suited to our task, since although creation and deletion are relatively expensive, they are also relatively infrequent, whilst context switching and function call and return are very cheap and will occur very often. The obvious alternative of spaghetti stacks [Bobrow and Wegbreit, 1972] was rejected because the implementation is very complex and personal experience of one of our team with the LISP/VM system [Alberga *et al.*, 1986] has shown that it is not so successful as it is widely believed to be.

4.2. Multiple Environment

The other implementation issue regarding multiprocessing in LISP concerns the maintenance and retention of the evaluation environment. The point at issue is handling free variable references. Because REDUCE is written in a traditional LISP, we have to maintain multiple dynamic binding environments - essentially, we have the classical funarg problem. There is a conceptually simple model (deep binding) which suffers from relatively slow identifier access ($O(n)$) but fast context switch ($O(1)$); the more complex model (shallow binding with rerooting), has $O(1)$ identifier access but $O(n)$ context switch. Since in a multi-processing environment we need both operations to be fast and because one of the team has spent a considerable amount of time investigating this problem, we have adopted a new model described in Padget and Fitch [1985] and in detail in Padget [1984]. Although the worst case complexity in this scheme for identifier access is no better than for deep binding, the constant factor is much smaller and context switching is $O(1)$; observed performance is very satisfactory.

The PSL system we use has been extended with stack groups and the Padget binding mechanism, and is running successfully on SUN-3 and HP Bobcats.

4.3. Parallel Tracing

While considering the parallel LISP, it is worth mentioning that in order to determine the dynamic behaviour of all the processors and tasks we have been forced to develop a graphical profiling tool, which is capable of showing time and store requirements of various parts of the system. This tool is described in detail in Burdoff *et al.*[1989].

4.4. Data and Programs

We have chosen to replicate most programs in all processors, with sharing being achieved by virtue of the multiprocessing LISP. At some later stage we will have to consider in more detail load balancing, and dynamic processes migration.

As we are committed to non-shared memory, there are a number of important issues in passing data between the processors. What is needed is a mechanism for building isomorphs of local structures on remote machines. What makes this problematic is the need to preserve shared items in the structure, which requires the copying process to keep track of the items it has seen in case they occur again. This changes the process from one linear in the depth of the structure to one linear in the size of the structure in a naïve implementation. Solutions are:

α to use mark bits on the structure in a manner analogous to garbage collection with auxiliary store proportional to the number of shared items in the structure, but this requires a second pass over the structure to remove the mark bits

β in a two space copying system, the alternate heap can be used to build a table of the shared items and the references to them, so this is essentially free.

There are also opportunities for using various data compression techniques in the transmission of data, and using the fast loading mechanism of LISP. We have made some small advances in this area, but there are still great opportunities for more.

In the case of large structures, when they do occur, it seems wasteful to spend a lot of time processing, converting and transmitting for only a small part to be used - naturally, this dynamic behaviour cannot necessarily be predicted. This suggests that a lazy-copying strategy might be worthwhile if the overhead was small enough.

5. CONCLUSIONS

At present we have only laid down ground work for a compilation driven parallel algebra system. It remains to be seen the extent to which we can evaluate REDUCE in parallel.

This paper has presented some aspects of the Intelligent Compilation approach to coarse-grained parallelism, and in particular the multiprocessor LISP system constructed from standard hardware, 68000-based work stations and Ethernet. This approach has formed part of the concurrent advanced simulation system at RAND, but it is more than that, being a class of methods which have a great potential for moderate sized assemblies of powerful processors. Our approach includes pragmatic aspects, concerned not only with where there is a scope for parallelism, but also where it is worthwhile.

6. REFERENCES

Alberga, C. N., Bosman-Clark, C., Mikelsons, M., Van Deusen, M. S. and Padget, J. A. (1986). Proc. ACM Sym. LISP and Functional Programmimg,

39-53.

Bobrow, D. G. and Wegbreit, B. (1972). **Comm. ACM, 15** 591-603.

Bowyer, A., Willis, P. J., and Woodwark, J. R. (1981). **Computer Journal, 24** 353-357.

Burdoff, C., Fitch, J. P., Marti., J. B. and Padget, J. A. (1989). *A Multiprocessor Profiling Tool*, Hawaii Systems Conference.

Fitch, J. P. and Norman, A. C. (1977). **Software - Practice and Experience, 7**, 713-725.

Fitch, J. P. (1986). *Notes on the Data Flow Analyser*, RAND CPAS internal document.

Griss, M. L., Bensen, E. and Maguire, G. Q. (1982). Proc. ACM Sym. Lisp and Functional Programming, 88-97.

Halstead, R.H. (1985). **ACM TOPLAS, 7**, 501-538.

Marti, J. B. (1980a). Ph.D. Dissertation, University of Utah.

Marti, J. B. (1980b). Proc. ACM Sym. Lisp and Functional Programming, 203-207.

Marti, J. B. and Fitch, J. P. (1983). Proc. EUROCAL 83, **LNCS 174**, Springer-Verlag, 78-90.

Marti, J. B. and Fitch, J. P. *Global Data Flow*, in preparation.

Padget, J. A. (1984). Ph.D. Dissertation, University of Bath.

Padget, J. A. and Fitch, J. P. (1985). Proc. 12th ACM Conference on Principles of Programming Languages, 255-265.

Weinreb, D. E. and Moon, D. (1984). *The Lisp Machine Manual*, Symbolics Inc.

PAC

TOWARDS A PARALLEL COMPUTER ALGEBRA CO-PROCESSOR

Jean-Louis ROCH[1]

Laboratoire TIM3 - IMAG
Institut National Polytechnique de Grenoble
46, avenue Félix Viallet
38031 Grenoble Cedex France

Abstract :

PAC is a computer algebra system, based on MIMD type parallelism. It uses parallelism as a tool for processing problems which are too complex for a sequential treatment. Basic fundamentals of the system are firstly discussed. Then, the implementation of infinite-precision arithmetic with vectors is detailed (addition, subtraction, multiplication, division in $\mathbb{Q}$, gcd and Bezout algorithm in $\mathbb{Z}$).

PAC is implemented on the Floating Point System hypercube : Tesseract 20 (16 nodes). To increase the speed of basic routines, some modules have been rewritten in assembly language for Inmos T414 transputer. Different experiments have been done to compare PAC perfomances (for basic arithmetic) to Maclisp.

---

[1] 1 This work is supported by the *PRC mathématiques et informatique* and by the *GRECO calcul formel* of the french *Centre National de la Recherche Scientifique* (CNRS).

COMPUTER ALGEBRA AND PARALLELISM ISBN 0-12-221100-6

INTRODUCTION

Each Computer Algebra system is composed of two distinct parts. On one hand, it includes an informatic part, managing symbols, rewrite-rules, patterns, abstract types handling.... On another hand, it contains many mathematical tools. Unless mathematical tools are carefully implemented, they cannot be used for solving problems of large size, cause of their too big complexities. For instance, under Macsyma, it is quasi-impossible to inverse a linear system with rationnal coefficients of size 100x100 (1). However, one of the credibility criteria of a system is its capacity to solve such big problems.

For those large problems, time spent in symbolic manipulations is negligible in respect with time spent in mathematical calculus (cf 1). Also, it seems interesting to use parallelism, because it allows to consider many problems under a new approach, decreasing in most cases their complexity.

In the present paper, we describe the implementation of a Computer Algebra system designed on a parallel MIMD machine. Many problems are met while conceiving such a system :

- For a given problem, the choice of the basic topological structure (hypercube, torus, ring...) is fundamental, and influences consequently complexity. This shows the necessity of developping basic routines, that do not depend on topologies adopted by the user to process different algorithms.
- To evaluate efficiency of algorithms, complexities have to be precisely studied : the sequential/parallel times comparison is often insufficient.
- Most of the existing systems allow a very convenient interactive use. How is such a mode possible, on a parallel machine ?

It is firstly essential to precise what type of parallelism (and so what machine) is well suited to problems we want to solve.

This study is followed in a general framework : we are implementing PAC on a particular computer, but this paper deals with problems of parallelism, and is a first description of a big computer algebra dedicated machine - this will be subsequently more precisely studied -.

I / PAC : BASIC PRINCIPLES

I.1 / Adequation of Parallelism to Computer Algebra

One of the basic principles of a Computer Algebra System is to allow building of embedded domains (in mathematical means). Thus, once the integer group $\mathbb{Z}$ has been built, the rationnal field $\mathbb{Q}$ or the integer coefficients polynomial ring $\mathbb{Z}[X]$ may be constructed, and after the fraction field $\mathbb{Q}(X)$. Those basic set extensions lead us to define bigger and bigger domains, whose elements have complex structures, built from already defined stuctures.

This constraint brings precisions about the type of parallelism the best suited to the problem.

In effect, in a first approach, two posibilities may be considered :

- Parallelism with low level nodes : we consider a processor network where each processor is able to communicate efficiently data, but can process only elementary operations. Such a structure is well suited to process "simple" algorithms, i.e. algorithms that don't work with complex structures. This approach is not very interesting for most of Computer Algebra algorithms, where parallelism may appear at a high level.
- Parallelism with high level nodes : we consider a processor network, where each processor gets a big memory area, with a good ability to process efficiently calculus. For example, if polynomials greatest common divisor "gcd" is to be processed, a good parallel way is spliting up modularly the problem on different values :

- Let us suppose polynomials are known by each processor
- Evaluating the polynomials in different points (one point, one processor)

- Performing the gcd calculation on the values so found (one integer gcd calculation one processor)
- Interpolating the different gcds obtained, to find the polynomial result. (on all processors)

Such an architecture (parallel with high level nodes) is also very well suited to process big elementary tasks. In the Computer Algebra field, it allows development of algorithms independently of the conditioning (essentially integers size) of the problem.

Moreover, most of algebraic operations in "small" finite fields (matrix or vector products, for example) may be very fast on a vector unit. So, disposing of a high rate pipelined arithmetic unit is fundamental. Besides, it allows an efficient and original processing of basic arithmetical operations. This will be subsequently followed and developped.

Those different criterions (high level nodes parallelism on one hand, and vector process unit (VPU) on each node on another hand) lead us to choose the Floating Point System (FPS) Tesseract 20 computer (16 processors connected according to an hypercube topology; each node is based on a INMOS T414 transputer with a 4 Mbytes memory, and contains a 16 Mflops VPU), on which we are implementing PAC [5].

I.2 / General Presentation of PAC

We are mainly interested in the parallelization of mathematical Computer Algebra algorithms : that's why the symbolic part is performed by a classical system (CAS) wich is installed on a host machine. This system may call parallel algorithms developped on a MIMD machine, directly connected to the host. Conversions between the different standards - on host and on the parallel machine - are performed by an interface. PAC may be also considered as an algebraic (soft) co-processor, associated to a Computer Algbra host system.

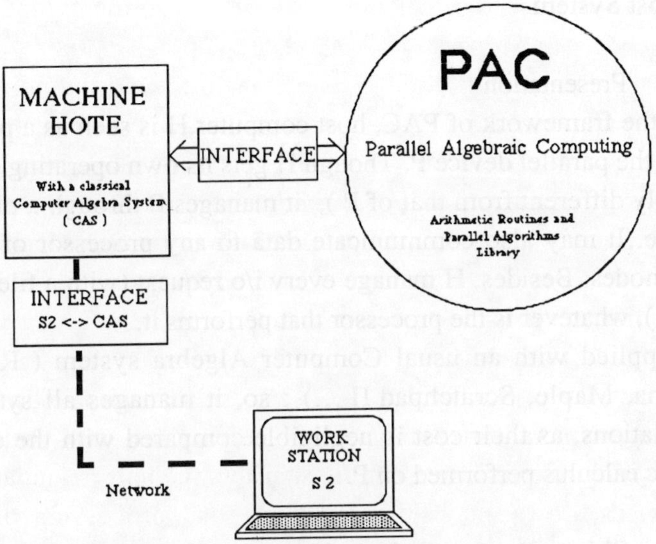

Figure 1 : General Overview of PAC

The first aim of PAC system is also to allow implementation of parallel algorithms, solving usual Computer Algebra problems on a MIMD machine, as considering the parallel machine as a peculiar device specialized in algebraic operations processing.

In the following, we will describe :

- H : the host computer
- CAS : the host Computer Algebra system
- P : the parallel machine connected to H

So, within the framework of the laboratory and of this study, H is a μVax II with Ultrix 2.1 (Unix 4.2),and P is the MIMD computer FPS Tesseract 20 (16 processors).

This approach may be generalized : P may be considered as a big device accessible by any Computer Algebra system via a network. P would be used for large tasks that can't be sequentially solved. The interface is then seen as a symbolic server.

I.3 / Host System

I.3.a / Presentation

Within the framework of PAC, host computer H is seen as a peculiar node of the parallel device P. Though H gets its own operating system (possibly different from that of P), it manages P through a physical interface. It may also communicate data to any processor of P, via system nodes. Besides, H manage every i/o request (with a file or the console), whatever is the processor that performs it.

H is supplied with an usual Computer Algebra system (Reduce, Macsyma, Maple, Scratchpad II ...) : so, it manages all symbolic manipulations, as their cost is negligible compared with the cost of algebraic calculus performed on P.

I.3.b / Objects Representation - Necessity of an interface

CAS implemented on H gets its own internal objects representation (polynomials, matrices, ...) more suitable for manipulations than external infix form. For the internal representation on P, we have to look at the different points :

- physical basic data representation (integers, reals, characters...) is often different on H and on P. Also, within the framework of the FPS T-20, H is a μVax II (standard DEC), although basic standard on P is IEEE.

- the parallel algorithms studied are very tied to the representation of objects, and especially to their sharing in memory between the different nodes of the hypercube. Also, let us consider as an example the product of two polynomials : two representations may be considered, each corresponding to a specific parallel algorithm :

- a sparse representation, allowing the distribution of a polynomial with respect to its monomials. This distribution corresponds to a block multiplication algorithm on each node.

- a dense recursive compact representation, allowing to perform easily polynomials evaluations with respect to one indeterminate. The polynomial product may also be obtained by modular calculus from results obtained on different processors.

So, a data translation phase is necessary, to communicate an object from H to P, and conversely. This phase is performed on H . When a parallel routine is called from H, the interface manages representation translations and memory distribution of input objects. Then, it calls the parallel P-routine, and it translates the result obtained on P with respect to the objects representation on H.

I.4 / Parallel Device System

Basic routines are implemented on the parallel device P. They contain an infinite-precision arithmetic (calculations in $\mathbb{Z}$ or $\mathbb{Q}$), and a polynomial and matrix arithmetic (implementation in progress).

The P-system constitues also a parallel routine library, allowing :

- resolutions of usual Computer Algebra problems (polynomial and matrix algebra)
- developments of parallel algorithms, using integer, polynomial, matrix... structures, with computation and communication primitive routines integrated in PAC.

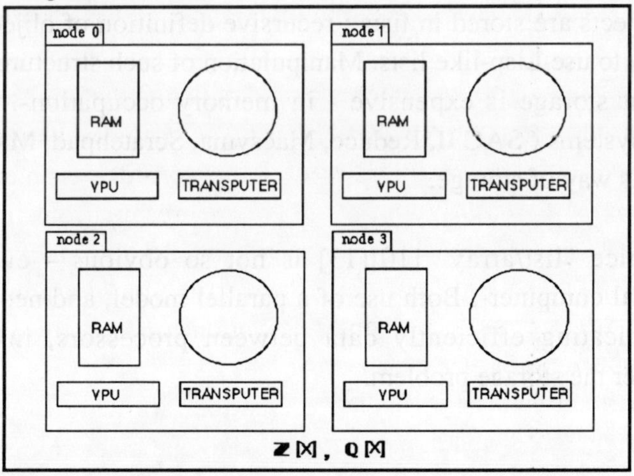

Figure 2 : Calculus Domains Distribution On the Parallel Device

I.5 / PAC Implementation

In the Grenoble TIM3 laboratory, PAC is developped on the FPS Tesseract 20. This parallel device is managed by a μVax II. On this host, MAPLE Computer Algebra System has been chosen as host system : it performs interactions user/system. The choice of MAPLE is essentially justified for its compatibility with C-programs.

A C-interface binds MAPLE to the parallel machine.

II / BASIC NODAL STRUCTURES

II.1 / Objects manipulated in Computer Algebra

One of the basic principles of Computer Algebra is the building of embedded domains (§I.1). For such recursive structures implementation, two structures are mainly used :

- objects are stored in arrays : this type of storage brings an efficient way to save up memory place. However, object manipulations are often expensive, and access to fields, inside an object, data dependents (Altran). One of the major problems is the impossibility of knowing exactly object size, a priori (What place is needed to store the product of 2 polynomials in ·[X], knowing the two polynomial operands, but without computing explicitely their product ???).
- objects are stored in lists : recursive definition of objects leads naturally to use Lisp-like lists. Manipulation of such structures is then easy, but storage is expensive - in memory occupation-. Most of current systems (SAC II, Reduce, Macsyma, Scratchpad, Maple) are using this way of storage.

The choice "list/array" [10][13] is not so obvious - even on a sequential computer-. Both use of a parallel model, and necessity of communicating efficiently data between processors, involve to reconsider the storage problem.

II.2 / Inherent constraints in the parallel model

Structures chosen for objects - possibly recursive - representation, have to respect parallel constraints - more specifically, within the framework of the FPS T20 -.

Those constraints are mainly of two orders :

- Parallelism constraints :

On the parallel device FPS-T20, time to communicate n 32 bits-words between 2 neighbouring nodes is [15]:

$$T_{com} = \beta + n\tau$$

$$\text{with } \beta = 9.5\ \mu s \quad \text{and} \quad \tau = 6\ \mu s$$

To communicate efficiently big data (e.g.polynomials), it would be better to store them as vectors, occupying adjacent places in memory.

However, depending on the chosen parallel type of algorithms, it is obvious that nodal manipulations are easier when using lists.

Let us consider as example, the product of two integer matrices. A Divide & Conquer approach of this problem leads to divide matrices in blocks : each block is constituted with integers stored adjacently in memory, so to speed up block communication time. However, when collecting the different results on the node which manages sharing, the complete matrix result should be stored as a bi-dimensionnal list. Important data communication time may so be speed up by using two types of structures - lists and vectors - simultaneously on each node.

- Constraints inherent in using vector process unit [9]:

Computation with a vector process unit is possible only if the both operands of a vectorial operations are stored in arrays, in two distinct banks. This is usually verified, whatever the VPU is, and that is also true concerning the FPS-T20 VPUs. This brings to store data contiguously in memory, in variable size vectors.

We are so led to the same conclusion as in the parallel constraint : to increase efficiency, we have to store objects in vectors. But simultaneously, we may use lists, to gather different vectors for the building of a greater domain, while keeping reasonnable access time to fields of the new structure.

II.3 / Variable size vectors :

It is so fundamental being able to manipulate vector structured data, occupying contiguous places in memory. In the following, we design as *block* such data.

In fact, a block is an array of a certain size, allocated in memory. To perform allocations and free blocks re-use, a garbage collector manages memory, processing memory merges and compactions[1][2]

Different contraints are closed to the manipulation of such objects :

- storing objects in memory in "constant" time, whatever the size is
- manipulating pointers inside blocks
- communicating efficiently blocks between nodes

Blocks Typing

To represent any object (for example, a matrix with coefficients in $\mathbb{Z}$, composed with contiguous memory blocks), new types have to be defined. All objects belonging to the same type, have a similar structure. To allow object typing, 8 bits are reserved in the first word of a block. Different types are already pre-defined : they are given, with corresponding codes, in the above-mentionned figure.

For a given block type, the different fields have to be accessed in the same way, whatever the field values are. Thus, it is possible to use pointers inside blocks.

Remarks

- In addition to the type zone stored in the first word of the block, 8 supplementary bits are usable :they can serve as extension of the block zone. This zone can be considered as the physical definition of sub-types. But it can also be used to store "short" field of a block.

- The definition of a new block can always be replaced by a list storage of such blocks, whose first element could - it is often the case in LISP - contain a type, and whose elements are blocks of already known types. But the communication times are always longer, and vectorization is impossible.

III / ARITHMETIC

III.1 / Positive integers domain : ℕ

Representation

An infinite precision positive integer is a typed block formed of the decomposition in 2^{32} basis of the represented number. This computation basis has been chosen because of the existence of an extended arithmetic on each transputer.

$$\text{Let } A \in \mathbb{N} \quad \text{such that} \quad A = \sum_{i=0}^{n} a_i\beta^i \quad \text{with} \quad \beta = 2^{32}$$

$$(0 \leq a_i < 2^{32} \text{ and } 0 < a_n < 2^{32})$$

It is stored under block representation.

Conversions - Inputs/Outputs

The interface user/internal representation necessitates 2 processes allowing the passage from the representation in 2^{32} basis to the representation in 10^9 basis - and reciprocally -. Primitive routines allows to read or to write an infinite precision integer from any node of the hypercube.

Arithmetic

The operations +,-,*,/,mod are implemented on each transputer, using classical algorithms [8][12]. Gcd' computation is realised using Lehmer algorithm, wich is very suitable to transputer [7].

Studies have been performed for the vectorization of some operations (+ and *). The storage of integers in block is then fundamental. The problem is in the carry propagation during the computation. The passage to a redundant notation would allow to avoid this propagation [3][11].

- Addition : on the adder, 4 steps :

1/ we compute r with $r_i = a_i + b_i$

2/ we compute c_i, if $r_i \geq b$ $c_i = 1$ else 0

3/ we compute r' with $r'_i = a_i + b_i + 1$

4/ we compute c'_i , if $r'_i \geq b$ $c'_i = 1$ else 0

5/ the results are then selected on the Transputer : we obtain R = a + b with : $R_i = (\mu_i * r_i + u_i * r'_i) \bmod b$ $i = 0..n$

with $u_0 = 1$ and $u_i = c_i \wedge u_{i-1} \vee c'_i \wedge \mu_{i-1}$ $i = 1..n$

- Multiplication : operations are chained on the adder and on the mutiplier :

To avoid carry propagation, we choose a basis less than the machine precision. The operations are chained until precision overflow. Let us assume that there are 53 bits of precision (calculus in 254 basis) on the vectorial unit, constituted of pipelined adder and multiplier.

We want to compute w =a*b+c with :

1/ passage from the 2^{32} basis (transputer) to the 2^{16} basis (Vector Processing Unit)

2/ Initialization : $w_i \leftarrow c_i$

3/ while <there remain some products to compute> do :

- calculus of $w_i \leftarrow a_k * b_{i-k} + w_i$
- conversion of w in 2^{32} basis

4/ translation of the result w

Obviously this algorithm does not minimize the performed operations (choice of a small computation basis). However, it is well suitable for the vectorization: 2^{22} additions and multiplications can be chained, with neither carries nor precision problems.

III.2 / Integer domain : $\mathbb{Z}$

Representation

An infinite precision integer is a typed block formed of the decomposition in 2^{32} basis of the absolute value of the represented number, together with a sign.

Arithmetic

The operations +,-,*,/,gcd, and conversions, are defined from the operations on the positive integers. It consists only in a study of the operands and result sign.

III.3 / Rationnal domain : $\mathbb{Q}$

Representation

An infinite precision rationnal is a typed block formed of the juxtaposition of the decomposition in 2^{32} basis of the numerator and of the denominator (The gcd computation and the reduction are not systematic for a given fraction : one bit indicates if the rationnal is reduced or not. However, special operations allow to compute with only reduced rationnals). Contiguous representation has been chosen to allow fast communications.

Arithmetic

Operations +,-,*,/,gcd, and conversions on $\mathbb{Q}$ are defined using operations defined earlier on $\mathbb{Z}$. They consist only in studying signs of operands and computations on numerators and denominators of operands, to set up the final result.

III.4 / Performances

It's very difficult to have a good evaluation of efficiency of arithmetic operations, as there is no other infinite precision arithmetic system implemented on transputers. We chose to compare PAC performances to Maclisp. We use the version of Maclisp implented on Bull-DPS 8 (Multics system). We compare times of compiled (C) and interpreted (I) Lisp programs.

Those diagrams show that we obtain the same order as Maclisp for complexity, with a certain factor of proportion (12). This remark is right for all arithmetic operations, except for division - this operation hasn't been yet optimize in assembly language-. But, comparizons time in Pac is very often constant (comparing the sizes and first words of both integers is often sufficient).

In fact, as the following diagrams proves it, this proportion factor comes essentially for the different capacities of the two machines : the T414 Transputer is based on a RISC architecture, while DPS8 is a CISC. Thus, as an example, it's impossible to use registers with the T414 (only the 4 registers A B C D).

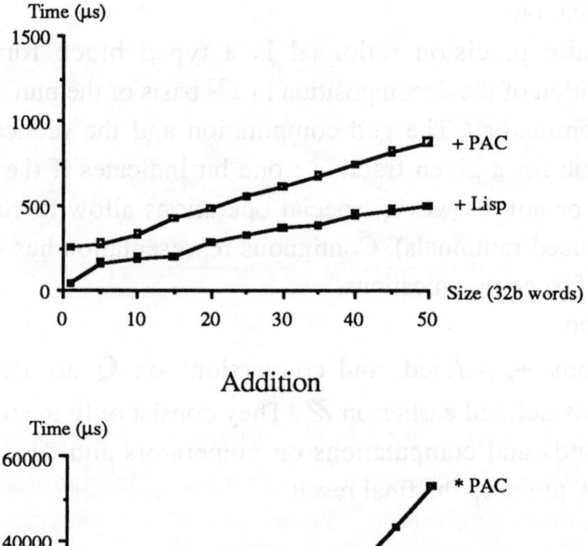

Addition

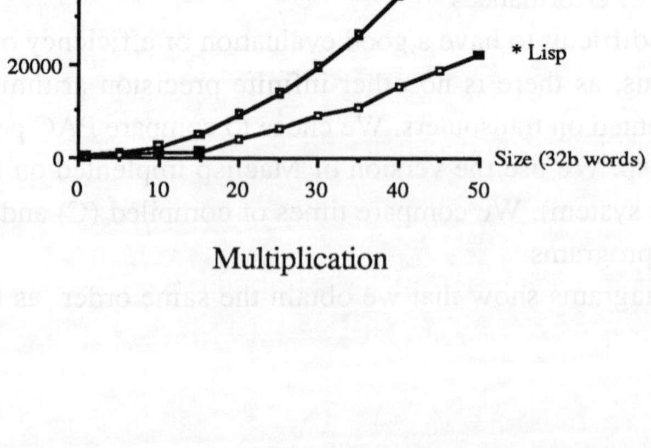

Multiplication

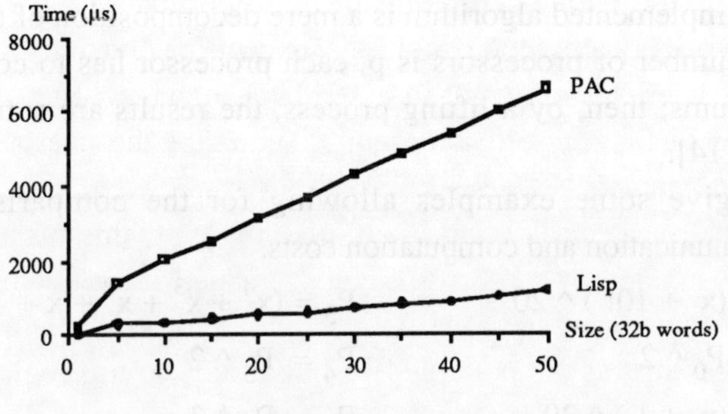

Division

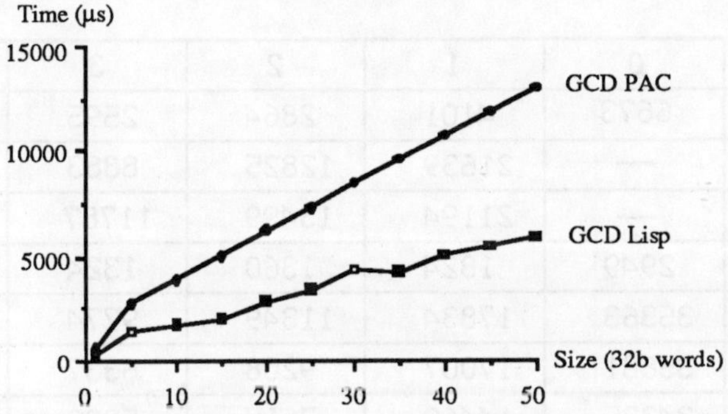

Greatest Common Divisor

IV / APPLICATION : PARALLEL MULTIPLICATION IN $\mathbb{Z}[X]$

We test the performance of parallel implementation on the multiplication of univariate polynomials - which is a typical problem of computer algebra - [4][6].

$$\text{Let } P = \sum_{i=0}^{n} p_i X^{e_i} \quad \text{and} \quad Q = \sum_{j=0}^{m} q_j X^{f_j}$$

If we consider the product as a sum of n+1 polynomials:

$$\sum_{i=0}^{n} p_i X^{e_i} * \sum_{j=0}^{m} q_j X^{f_j} = \sum_{i=0}^{n} \left(\sum_{j=0}^{m} p_i q_j X^{e_i + f_j}\right)$$

The implemented algorithm is a mere decomposition of the n sums. If the number of processors is p, each processor has to compute about n/p sums; then, by a lifting process, the results are summed two by two [14].

We give some examples allowing for the comparison between communication and computation costs.

$P_0 = (x + 10!)\ \hat{}\ 20$ $\qquad$ $P_3 = (x^4 + x^3 + x^2 + x + 1)\ \hat{}\ 5$

$P_1 = P_0\ \hat{}\ 2$ $\qquad$ $P_4 = P_3\ \hat{}\ 2$

$P_2 = (x + 1)\ \hat{}\ 20$ $\qquad$ $P_5 = P_3\ \hat{}\ 3$

	0	1	2	3	4
$P_0 * P_0$	6673	4101	2864	2595	2693
$P_1 * P_0$	—	21639	12825	8883	7881
$P_0 * P_1$	—	21194	13499	11787	11778
$P_2 * P_2$	2949	1824	1360	1324	1465
$P_3 * P_5$	35363	17834	11349	9774	9804
$P_5 * P_3$	35801	17007	9208	6337	5570
$P_4 * P_4$	34325	14460	7610	5300	4833

Time in ms against hypercube dimension

For examples 2 and 3, on one processor, the coefficients size leads to memory overflow. From a certain number of processors, communication cost becomes too expensive with respect to the arithmetic costs; to obtain good performances, the calculus to perform has to be consequent in each processor.

It appears that for a sufficient number of processors it is more efficient to multiply the highest degree polynomial by the other one (more communication time) [see ex. 2-3 and 5-6].

V / CURRENT STATE OF THE PROJECT

Several parallel algorithms of computer algebra have been implemented in PAC. They concern essentially linear algebra problems (Diophantine equations (F.Siebert), linear systems (G.Villard)) and some problems of pure algebra (Grobner basis (P.Senechaud)).
The interface problems with Maple are not yet been solved.
Several complexity studies are done on the forme problems ([14]), and point out the parallelism efficiency as soon as the problems size becomes important.

REFERENCES

[1] A.V. Aho, J.E. Hopcroft, J.D. Ullman (1983), Data Structure & Algorithm, Addison-Wesley

[2] Harold Abelson, Gerald Jay Sussman, Julie Sussman (1985), Structure & Interpretation of Computer Programs, Mc Graw-Hill Book Company, pp. 491-503

[3] A.Avizienis (1961), Signed-digit Number Representations for Fast Parallel Arithmetic, IRE Transactions on Electronic Computers

[4] G.E.Collins, PM, A System for Polynomial Manipulation, ACM, vol.9, n°8

[5] M. Cosnard, B. Tourancheau, G. Villard (1987, Présentation de l'hypercube T20 de FPS, Revue Bigre + Globule

[6] S.C.Johnson, Sparse Polynomial Arithmetic, Research report, Bell Telephon Labs, Murray-Hill, N.J.

[7] D.H.Lehmer (1938), Euclid's Algorithm for Large Numbers, AMM 45, pp 227-233

[8] D.E. Knuth (1969), The Art of Computer Programming, Vol. 2, Semi-Numerical Algorithms, Addison-Wesley Reading Mass, pp. 229-293

[9] P.M.Kogge (1980), The Architecture of Pipelined Computers, Mc Graw-Hill

[10] Robert Mœnck, Is a Linked List the Best Storage for an Algebra System, Research Report

[11] J.M. Muller (1988), Arithmétique des ordinateurs, Masson - to appear -

[12] E. Regener (1984), Multiprecision Integer Division Examples Using Arbitrary Radix, ACM, vol. 10 N° 3

[13] J.L. Roch, P. Senechaud, F. Siebert, G. Villard (1987), Calcul Formel, Parallelisme et Occam, OPPT Ed T.Muntean

[14] J.L. Roch, P. Senechaud, F. Siebert, G. Villard (1987) Parallel Algebraic Computation, Rapport de Recherche Imag/RR686-I-

[15] Y. Saad and M. Schultz (1985), Topological Properties of Hypercubes Research Report YaleU/DCS/RR-389

Evaluation of "Performance Enhancements" in Algebraic Manipulation Systems (thesis abstract)

Carl G. Ponder

Computer Science Division
University of California, Berkeley
Berkeley, CA 94720

September 26, 1988

Abstract

This paper is an abstract of my thesis examining several proposed ways to speed up symbolic algebraic computation: hashing techniques, parallel processing, application of the *FFT*, and alternative representations of polynomial expressions.

Polynomials, and variants like rational functions, truncated power series, and Poisson series, represent an important class of expressions in algebraic manipulation. Parallel algorithms are analyzed for multiplying and powering sparse and dense polynomials, including parallel forms of the *FFT*. Alternative polynomial representations are compared and it is suggested that an efficient algebraic manipulation system might use a family of polynomial representations rather than one general form.

Gröbner-basis reduction is an inherently hard problem, yet can be used as a powerful tool in computational mathematics. Contrary to recent claims, empirical studies show that it is difficult to exploit parallelism in Gröbner-basis computation.

A variety of hashing mechanisms have been proposed for performing operations in symbolic computation, including *memo functions* for recognizing and eliminating redundant computations. The proposed mechanisms are presented in summary; empirical study shows that memo functions, except in certain circumstances, may not be an advantage.

Additional summary of relevant results in parallel algorithm design and parallel processing machines and languages is presented.

COMPUTER ALGEBRA AND PARALLELISM ISBN 0-12-221100-6

Motivation and Preview

Algorithms in symbolic algebra systems range from polynomial-time (matrix inversion) to superexponential-space (Gröbner-basis reduction). A number of programs attempt to solve problems that are in general undecidable (equivalence of expressions). In practice, solution to useful problems can easily take hours or days to produce. The irony is that some of these problems can be expressed quite simply. In problems like factoring of polynomials or Gröbner-basis reduction, it is not unusual for small expressions to produce huge intermediate expressions even when the final result is small. The practical expense of operations is quite apparent to interactive users of systems like Macsyma. Even as processors become faster and memory less expensive, investigations continue to pose more time-consuming problems. Additional speedups due to parallel execution as well as better algorithms or representations provide real benefits.

In this thesis we look at some possible methods for speeding up algebraic computations. These involve parallel algorithms, alternative representations, and various hashing mechanisms. A combination of formal and empirical analyses are used in evaluating these ideas. Some summaries of relevant work in various fields are presented both to identify the state of the art (which seems rather primitive, from a practical standpoint) and to suggest areas where future work should progress.

Chapters 2 and 3 survey current results in parallel algorithm design and the implementation of parallel languages and systems relevant to algebraic manipulation. In summary, few *efficient* parallel algorithms have been formulated or analyzed for significant problems, which is hardly reproachable since the serial complexity of many significant problems is poorly understood. Gröbner-basis reduction [3] is one such poorly understood problem, which happens to be important since it can be used to solve a wide variety of problems in algebraic geometry. Several "obvious" parallel algorithms for Gröbner-basis reduction are tested in chapter 7, none appearing to be particularly useful.

Numerous parallel languages and architectures have been proposed up to now; some rather poor experiences with the "state of the art" (Alliant Qlisp and Intel Hypercube Concurrent Common Lisp) are reported in chapters 3 and 5. The dream of constructing an efficient parallel system for algebraic manipulation is still out of reach; in the meantime it is difficult to realistically test any parallel algorithms, much less produce any useful software.

Chapters 4, 5, and 6 deal with the general topic of polynomial manipulation. Chapter 4 discusses the issues involved with specific polynomial representations, and suggests an efficient combination of representations. Chapters 5 and 6 analyze parallel algorithms for multiplying and powering sparse and dense polynomials, using such algorithms as the FFT. Formal analysis suggests a linear speedup for a fixed number of processors and large inputs. Algorithms for multiplying sparse polynomials are evaluated empirically with positive results, ignoring the poor quality of the Alliant Qlisp testbed. An efficient parallel system for algebraic manipulation may well include parallel routines for multiplying and powering polynomials, to be used automatically or by user choice given sufficiently large polynomials.

Chapters 8 and 9 deal with hashing mechanisms for more efficient symbolic manipulation. Chapter 8 surveys proposed techniques, including the possibility of efficient pattern-matching for nonrandom data. Chapter 9 analyzes the "memo function" mechanism of the Maple system [5], which tabulates the input and output of a given set of functions to eliminate recomputation. Empirical analysis inside the Macsyma system does not favor the use of memo functions, although some of the Maple design decisions may make memo functions useful. Unfortunately the performance of Maple has not been analyzed sufficiently to resolve this question.

This thesis is rather loosely organized, treating each topic in a distinct chapter. The chapters have been submitted to various publications as disjoint papers, the titles for chapters 2-9 being *Parallelism and Algorithms for Algebraic Manipulation: Current Work*, *Parallel Processors and Systems for Algebraic Manipulation: Current Work*, *Polynomial Forms and Representations for Algebraic Manipulation*, *Parallel Multiplication and Powering of Sparse Polynomials*, *Parallel Multiplication of Dense Polynomials*, *Parallel Algorithms for Gröbner-Basis Reduction*, *Applications of Hashing in Algebraic Manipulation (an Annotated Bibliography)*, and *Augmenting Expensive Functions in Macsyma with Lookup Tables*. Chapters 2 and 3 have been submitted to *SIGSAM Bulletin;* chapter 8 has appeared in the November 1987 issue (21,4) of *SIGSAM Bulletin*. Chapter 7 has been submitted to *Transactions on Mathematical Software,* and chapters 5, 6, and 9 have been submitted to the *Journal of Symbolic Computation*. We believe the text of this thesis will be relatively consistent with whatever versions that might appear.

For the purposes of this workshop we will present only the results from chapters 5, 6, and 7 on polynomial operations and Gröbner-basis reduction; the survey material and the results on hashing are left to the full thesis, which is available as a tech report [14].

1 Parallel Polynomial Operations

Polynomials represent an important class of expressions in algebraic manipulation. Polynomials can be (crudely) divided into two classes, *sparse* and *dense*. In the sum-of-products expansion of a polynomial, if "most" of the possible terms for the given degree are nonzero, the polynomial is *dense*, otherwise it is *sparse*.

Multiplication and powering are frequently performed operations on polynomials. Here we will show algorithms which are asymptotically optimal (with regard to monomial multiplications) for multiplying and powering sparse polynomials in expanded form, called *Simp* and *NOMC*. The *FFT* algorithm is used for multiplying and powering dense polynomials, and is conjectured to be optimal.

The *Simp, NOMC,* and *FFT* algorithms can be parallelized "perfectly" with regard to monomial multiplications, gaining a linear reduction in the number of parallel steps for suitable numbers of processors. Both *Simp* and *NOMC* suffer in regard to the ordering operations used to collapse identical terms together. We expect the cost of monomial multiplication to dominate the cost of other operations to such a high degree, that for fixed numbers of processors and sufficiently large input, the serialization of the ordering operations will not significantly affect performance.

1.1 Characteristics of Polynomials

The following relations show how the degree and size of polynomials change under the operations of multiplication and powering:

$$\text{degree}(p_1p_2) = \text{degree}(p_1) + \text{degree}(p_2) \tag{1}$$

$$\text{size}(p_1p_2) \leq \text{size}(p_1) \cdot \text{size}(p_2) \tag{2}$$

$$\text{degree}(p < k) = k \cdot \text{degree}(p) \tag{3}$$

$$\text{size}(p < k) \leq \binom{\text{size}(p) + k - 1}{k} \tag{4}$$

as derived in [8]. This is approximated by

$$\text{size}(p < k) \sim \frac{\text{size}(p) < k}{k!} + O(\frac{\text{size}(p) < k - 1}{2(k-2)!})$$

for large k and increasing size(p). Letting v be the number of variables,

$$\text{size}(p) \leq \sum_{i=0} < v \binom{v}{i} \binom{\text{degree}(p)}{i}, \tag{5}$$

which reduces to

$$\text{size}(p) \leq \text{degree}(p) + 1 \tag{6}$$

for $v = 1$.

For completely dense polynomials relation (6) is a strict equality. Relations (2) and (4) become overly crude, with stronger bounds dictated by relations (1), (3), and (6).

Over a domain with zero-divisors, relation (1) is adjusted to

$$\text{degree}(p_1 p_2) \leq \text{degree}(p_1) + \text{degree}(p_2) \tag{1a}$$

Otherwise the degree will grow linearly under multiplication and powering. The size will grow at most quadratically (in the sparse case) and at least linearly (in the dense case, assuming no cancellation) under multiplication. Since size is ultimately bounded by degree, quadratic growth cannot be sustained under repeated multiplications as polynomials "fill in". As the "density" of the results increases, relatively fewer distinct terms will be generated by multiplication.

1.2 Complexities of Multiplication and Powering

We begin by showing that the number of scalar multiplications required for polynomial multiplication or powering depends strongly upon the size of the result.

Lemma 1:

At least $\text{size}(p_1 p_2) - \text{size}(p_1) - \text{size}(p_2)$ multiplications are required in the worst case to compute $p_1 p_2$.

Proof:

Let A, B, and C be the sets of coefficients in p_1, p_2, and $p_1 p_2$, respectively. If the elements of A and B are algebraically independent, the elements of C will be (at least) linearly independent. Given that we have generated C with h multiplications, let $E = \{e_1, ..., e_h\}$ be the set of products of the multiplications. C must be formed by linear combinations of A, B, and E:

$$c_j = \sum_{i=1} < h x_{i,j} e_i + \sum_{i=1} < n_1 y_{i,j} a_i + \sum_{i=1} < n_2 z_{i,j} b_i$$

$$1 \leq j \leq \text{size}(p_1 p_2)$$

> Since the elements of C are linearly independent, it must be the case that $\text{size}(p_1p_2) \le h + \text{size}(p_1) + \text{size}(p_2)$, or $h \ge \text{size}(p_1p_2) - \text{size}(p_1) - \text{size}(p_2)$. □

As a simple corollary, for p_1 and p_2 "sufficiently sparse", $\text{size}(p_1p_2) = \text{size}(p_1) \cdot \text{size}(p_2)$, so roughly $\text{size}(p_1) \cdot \text{size}(p_2)$ multiplies are required. The *Simp* algorithm operates in exactly this bound. For "sufficiently dense" polynomials, $\text{size}(p_1p_2) = \text{size}(p_1) + \text{size}(p_2)$ so the lower bound on multiplications is linear. The best known algorithm in this range is the *FFT*, using $\Theta((\text{degree}(p_1) + \text{degree}(p_2)) \cdot \log(\text{degree}(p_1) + \text{degree}(p_2)))$ scalar multiplications. For powering, we have the similar result [8]:

Lemma 2:

> At least $\text{size}(p < k) - \text{size}(p)$ scalar multiplications are required to compute $p < k$.

Proof:

> As before, the coefficient set C is linearly independent so the number h of intermediate products must satisfy the inequality $\text{size}(p < k) \le h + \text{size}(p)$, or $h \ge \text{size}(p < k) - \text{size}(p)$.

We know from relation (4) how the size of $p < k$ can grow for sparse polynomials. The algorithm *NOMC* operates asymptotically with this. For dense univariate polynomials the growth is bounded more strictly by relations (3) and (5), giving a bound of $k \cdot \text{size}(p)$. The best known algorithm for this case is the *FFT* using $\Theta(k \cdot \text{size}(p) \cdot \log(k \cdot \text{size}(p)))$.

Comparison operations are required to order the results. The problem of ordering the monomials of the product $p_1 \cdot p_2$ is equivalent to the "Sorting X+Y" problem (discussed by Harper *et. al* [11]) of ordering all pairwise sums of the (ordered) elements of two vectors. For $\text{size}(p_1) = \text{size}(p_2) = n$, the lower bound on the number of comparisons is $\Omega(n \lg n)$. A solution by Jean Vuillemin [15] uses $O(n < 2)$ comparisons, i.e. proportional to the largest possible size of the result. Ordering the monomials of $p < k$ is equivalent to sorting all distinct k-wise sums of elements of the vector X. Vuillemin's result is extended by recursively sorting the sets of $\lfloor \frac{k}{2} \rfloor$- and $\lceil \frac{k}{2} \rceil$- wise sums and sorting the pairwise sums of these two sets as before. The work to sort the final set will dominate the lower-order sets (see the analysis of *NOMC* in section (5)), so the total number of comparisons used is proportional to the largest possible size of the result

$$\binom{\text{size}(p)+k-1}{k}$$

Terms are combined by the addition operations, once comparisons have established that they are additive. Thus a lower bound on the number of coefficient additions required is

$$\text{size}(p_1)\cdot\text{size}(p_2)-\text{size}(p_1 p_2)$$

or

$$\binom{\text{size}(p)+k-1}{k}-\text{size}(p<k),$$

which is the number of terms possible under relations (2) or (4), minus the actual size of the result. In the worst case $p=p_1=p_2$ and are completely dense. Under relations (1), (3), and (6) the number of coefficient additions required are

$$\text{size}(p)<2-2\cdot\text{size}(p)$$

or

$$\binom{\text{size}(p)+k-1}{k}-k\cdot\text{size}(p)\sim\frac{\text{size}(p)<k}{k!}+k\cdot\text{size}(p)$$

for multiplication and powering, respectively. Assuming that the results are ordered, the coefficient additions amortize into the exponent comparisons. The addition operations involved in generating new exponents amortize into the coefficient multiplications.

It is not obvious how to parallelize the addition or comparison operations in either the *Simp* or *NOMC* algorithms. In each case terms generated by different processes may add together (or even cancel). In both the *Simp* and *NOMC* algorithms we will use a parallel mergesort [1] (just parallelize the recursive calls), which requires

$$\Theta(x+\frac{x}{y}\log\frac{x}{y})$$

comparisons to sort x items with y processors, y ranging from 1 to x. For fixed y and increasing x it approaches a speedup linear in y, though this is not a linear speedup over Vuillemin's result. A reasonably efficient parallel mergesort has been developed by Cole [6], which takes $\Theta(\log x)$ operations to sort x items with x processors. Parallelism is applied to operations on individual elements, a very fine level of granularity. This process still appears inefficient and unnecessarily complicated for (small) fixed numbers of processors and large inputs, particularly if there is a significant overhead to interprocessor communication or shared memory access.

A hash table of monomials can be used as an unordered representation of polynomials [10]. A hash table can be updated in parallel, or used in serial to reduce the number of comparisons required to combine terms. The integrity of a hash table is difficult to maintain under parallel updates. If this were not an issue, the operations involved with combining terms would parallelize perfectly. The details involved with locking the hash buckets are complicated enough to possibly negate any advantages. Additional overheads such as computing the hash function are also significant.

1.3 The *Simp* Algorithm for Multiplication

A simple way to multiply polynomials p_1 and p_2 is as follows [9]:

Given polynomials p_1, p_2, return $p_1 \cdot p_2$.

[1] If p_1 is a monomial, multiply each term of p_2 by p_1 and return result. Otherwise,

[2] Split p_1 into A and B, and recursively form the products $A \cdot p_2$ and $B \cdot p_2$.

[3] Merge the two partial products (ordered by exponent), adding coefficients with the same exponent. Return result.

This effectively decomposes polynomial p_1 into size(p_1) monomials, forms the product of each monomial with p_2, and successively merges the results. The parallelized form is to perform the recursive calls in step 2 in parallel.

The number of coefficient multiplications required is size(p_1)size(p_2). The merge step is a recursive balanced merge, requiring $O(\text{size}(p_1)\text{size}(p_2) \cdot \log(\text{size}(p_1)\text{size}(p_2)))$ comparisons. There are size(p_1)size(p_2) − size(p_1p_2) additions.

In parallel, $k \leq \text{size}(p_1)$ processors can be used. The recursive step 2 is better replaced by splitting p_1 k ways and having each processor perform the multiplication as before. Step 3 reduces to performing a parallel mergesort (section 3) on the partial products. The number of parallel multiplications is $\frac{1}{k}\text{size}(p_1)\text{size}(p_2)$, since multiplications are performed on the bottom level of the decomposition.

The parallel mergesort uses a parallel measure of comparisons $O(\text{size}(p_1p_2)+\frac{\text{size}(p_1p_2)}{k}\log\frac{\text{size}(p_1p_2)}{k})$, which is asymptotically $O(\text{size}(p_1p_2))$ as k approaches $\text{size}(p_1)$. Replacing the parallel merges with a (serial) k-way balanced merge gives a parallel measure of comparisons $O(\text{size}(p_1p_2)\cdot\log(k)+\frac{\text{size}(p_1p_2)}{k}\log\frac{\text{size}(p_1p_2)}{k})$. For message-passing multiprocessors with a high communication cost, this decomposition is probably superior.

Thus, given sufficient processors, the best time for the multiplications is $O(\text{size}(p_1))$ and for additions and comparisons is $O(\text{size}(p_1p_2))$. For $\text{size}(p_1) = \text{size}(p_2) = n$, this amounts to a reduction by $\Theta(n)$ in the time to perform multiplications and $\Theta(\lg(n))$ in the time to perform comparisons. In the worst case the number of additions is reduced at most by a constant factor, since $\Theta(n < 2)$ terms can combine in the final merge.

1.4 The *NOMC* Algorithm for Powering

The *NOMC* algorithm *(full multinomial expansion with dynamic programming)* is one of many asymptotically efficient algorithms for powering sparse polynomials. Several alternatives are mentioned in [8]. The algorithm is expressed as follows:

Given a polynomial $p = (a_1 + \ldots + a_t)$ and a power k to be computed, return $p < k$.

[1] If p is a monomial, power it and return the result (e.g. $p = cx_1 < i_1 \cdots x_n < i_n$, where c is a coefficient, return $p = c < kx_1 < i_1 + k \cdots x_n < i_n + k$).

[2] Tabulate products of powers of each a_n of total degree $\lfloor\frac{k}{2<i}\rfloor, \lceil\frac{k}{2<i}\rceil$ for $i = 0, \ldots, \log(k)$ using the relation
$a_1 < I_1a_2 < I_2 \cdots a_t < I_t = (a_1 < J_1a_2 < J_2 \cdots a_t < J_t) \cdot (a_1 < K_1a_2 < K_2 \cdots a_t < K_t)_{ii}$
$I_1 + \cdots + I_t = m, \quad J_1 + \cdots + J_t = \lfloor\frac{m}{2}\rfloor, \quad K_1 + \cdots + K_t = \lceil\frac{m}{2}\rceil.$

[3] Return $\sum_{I_1+\cdots+I_t=k}\binom{k}{I_1I_2\cdots I_t}a_1 < I_1a_2 < I_2 \cdots a_t < I_t, \quad I_1 + \cdots + I_t = k.$

For k a power of 2, the number of monomials tabulated is

$$\sum_{i=1} < \log(k) \binom{\text{size}(p) + \frac{k}{2<i} - 1}{\text{size}(p) - 1} < \binom{\text{size}(p) + k - 1}{\text{size}(p) - 1} + \sum_{i=1} < \frac{k}{2} \binom{\text{size}(p) + i - 1}{\text{size}(p) - 1}$$
$$= \binom{\text{size}(p) + k - 1}{\text{size}(p) - 1} + \binom{\text{size}(p) + \frac{k}{2}}{\text{size}(p) - 1} \sim \binom{\text{size}(p) + k - 1}{\text{size}(p) - 1}$$

Since each monomial is formed by multiplying two monomials from the table (using one coefficient multiply and v exponent additions), the number of coefficient multiplications is asymptotically the same as the number of monomials in the result. The higher-order term continues to dominate for k not a power of 2. Each successive multinomial coefficient is generated from a previous one using one integer multiply and up to one integer divide.

For a parallel implementation, the result monomials are broken into groups. Each processor fills in what is needed in the monomial table to compute its group of monomials. The coefficient multiplications are parallelized perfectly for up to

$$o \left[\frac{\binom{\text{size}(p) + k - 1}{\text{size}(p) - 1}}{\binom{\text{size}(p) + \frac{k}{2}}{\text{size}(p) - 1}} \right]$$

processors, since the overhead of constructing the lower-order monomials is dominated by the cost of generating the k-order monomials. For large k and increasing size(p), this bound approaches

$$\mathcal{K} = \frac{\text{size}(p) < k/2}{k \cdot (k-1) \cdot \cdots \cdot (k/2)}$$

processors. Beyond this, the second-order term

$$\binom{\text{size}(p) + \frac{k}{2}}{\text{size}(p) - 1}$$

becomes significant and restricts the asymptotic speedup. Therefore the best time for the multiplications is

$$\binom{\text{size}(p) + \frac{k}{2}}{\text{size}(p) - 1}$$

For K processors the parallel mergesort requires a number of comparisons asymptotically proportional to

$$\begin{pmatrix} \text{size}(p) + k - 1 \\ \text{size}(p) - 1 \end{pmatrix} \sim \frac{\text{size}(p) < k}{k!}$$

for large k and increasing size(p). Therefore the best time for the additions and comparisons is

$$O(\frac{\text{size}(p) < k}{k!}).$$

1.5 Empirical Results

An experiment was run to compare parallelized versions of the *Simp* and *Karatsuba* [9] algorithms, on a 4-processor Alliant running Qlisp. These were implemented as recursive algorithms with the *Simp* algorithm generating 2 processes per level of recursion and the *Karatsuba* algorithm generating 3. Each program we tested uses the parallelized algorithm up to a fixed number of processes, and the serial algorithm afterward. Table 1 shows the result in milliseconds, garbage-collection time excluded.

The *Karatsuba* algorithm achieved almost perfect linear speedup up to the number of actual processors available. The *Simp* algorithm didn't quite, but still ran strictly faster as processors were added, for this input. The cutoff point for *Karatsuba* to outrun *Simp* was between $(x < 60 + x < 59 + \cdots + x + 1) < 7$ and $(x < 80 + x < 59 + \cdots + x + 1) < 7$ (a well-coded *FFT* should outrun both for as low as $(x < 5 + x < 4 + \cdots + x + 1) < 3$); unfortunately the Qlisp system broke for problems that large.

A serial version of the powering problem ran in 1283 ms. on a VAX 11/785 with (Franz Inc.) Common Lisp and 3537 ms. on the Alliant with

Table 1 – Time to expand $(x < 13 + x < 12 + \cdots + x + 1) < 7$ with 4 processors.

# processes	*Simp*	*Karatsuba*
1	3537	7589
2	1864	——
3	——	2375
4	1067	——
8	1100	
9	——	1875

one-processor Qlisp; the fact that the Alliant is in other respects generally faster than the VAX suggests that Qlisp is poorly implemented.

1.6 The *FFT* Algorithm

The *FFT* algorithm [1] [2] [13] [17] is useful for multiplying dense polynomials with coefficients from the field of complex numbers or a finite computation structure (typically the integers modulo a prime). The precision or size of the modulus must be decided *a priori*, to be at least as large the precision of the result. This is an inconvenience in algebraic manipulation systems where arbitrary integer coefficients are commonly used.

The *FFT* works by computing the *discrete Fourier-Transform* of the univariate polynomial $p(x)$, which is the vector $[p(\omega < 0), ..., p(\omega < s-1)]$ where d is the degree of $p(x)$, the size of the transform $s \geq d+1$, and ω is some *principal sth root of unity*. The inverse of the transform is computed in a nearly identical way; effectively the transform vector can be regarded as the coefficients of the polynomial $p'(y) = \sum p(\omega < i) y < i$ and the vector of coefficients of $p(x)$ is $[\frac{p'(\omega<-0)}{n}, ..., \frac{p'(\omega<-s+1)}{n}]$. By evaluating two polynomials at $\omega < 0 ... \omega < s-1$, multiplying the corresponding values together, and converting the result back, the product polynomial is produced. Likewise, powering is performed by evaluating the polynomial, powering each resulting value, and converting back to get the powered polynomial.

The (Cooley-Tukey) *Fast Fourier Transform (FFT)* is the basic algorithm for computing the discrete Fourier transform in time $O(d \log d)$. It works as follows:

Let s be a power of 2, $s \geq d+1$
Let $A = [a_0, ..., a_{s-1}]$ be the coefficients of $p(x)$ (padded with zeros if necessary) in some computation structure C.
Let $A' = [a'_0, ..., a'_{s-1}]$ be the coefficients in the transformed polynomial $p'(y)$.
Let ω be a primitive sth root of unity in C.
$A' = FFT(A, s, \omega)$.

Recursive $FFT(A, s, \omega)$:
[1] if $s = 1$, return a_1.
[2] split coefficients by index into even-indexed sequence B and odd-indexed sequence C.

[3] $B' \leftarrow FFT(B, s/2, \omega < 2)$, $C' \leftarrow FFT(C, s/2, \omega < 2)$.
[4] for $i \leftarrow 0$ to $s/2 - 1$ do
$A'_i \leftarrow B'_i + \omega < iC'_i$.
$A'_{i+s/2} \leftarrow B'_i + \omega < i + s/2C'_i$.
[5] return A'.

Some multiplication and addition operations are hidden in the powering and the manipulation of the indices. Asymptotically, $\Theta(d \log d)$ additions and multiplications are performed since each step of the recursion performs work proportional to the subproblem size, and the recursion works on subproblems of exactly half the size. d must be rounded up to a power of two. Numerous tricks can be used to trim the number of operations [1] by a constant factor, specifically by "unraveling" the recursion and using bit-operations to permute the coefficients as necessary. An iterative form of the *FFT* can be stated as follows:

Iterative $FFT(A, s, \omega)$:
[1] for $l \leftarrow 0$ to $\lg(s)$ do
[2] for $i \leftarrow 0$ to $s - 1$ do $t_i \leftarrow A_{i+1}$.
[3] for $i \leftarrow 0$ to $s - 1$ do
$r_i \leftarrow t_{i \wedge (-2 < \lfloor \lg i \rfloor - l)} + \omega < \frac{s}{2 < l+1} \cdot bitreverse(l) \cdot t_{i \vee (2 < \lfloor \lg i \rfloor - l)}$.
[4] for $l \leftarrow 0$ until $s - 1$ do $A'_{bitreverse(l)} \leftarrow r_l$.
[5] return A'.

where $\lfloor \lg x \rfloor$ is the greatest integer no larger than the log base-2 of x, $bitreverse(x)$ is the reversal of the bits in x (within the fixed word length), and $\wedge$ and $\vee$ are bitwise *and* and *or*, respectively.

The *Good-Winograd* algorithm [2] [17] provides another decomposition for cases where s is *not* a power of 2, but is a product of two relatively prime integers of roughly equal size. The Good-Winograd algorithm factors the degree into two relatively-prime numbers and treating the polynomial as if it were bivariate. Powers of x are implicitly replaced by power-products of the two new variables. Either the Good-Winograd or the Cooley-Tukey algorithms can be applied to the subproblems, depending upon their size. The complexity is the same as the Cooley-Tukey algorithm.

1.7 Parallel Implementation of the *FFT*

Both the Cooley-Tukey and the Good-Winograd algorithms parallelize effectively. Step 3 of the recursive algorithm can be parallelized to give a parallel running time of $\Theta(d + \frac{d}{k}\log\frac{d}{k})$ for k processors, adding up to time $\Theta(d)$ for $k \in \Theta(d)$ processors. This decomposition is suitable for message-passing multiprocessors, where each process subdivides the problem and initiates a new process for each subproblem. The more intelligent approach would be to have one process divide the coefficient array k ways in step 2 and apply the serial iterative algorithm to the subprocesses. The combining step 4 can be performed recursively in parallel as before. The asymptotic complexity remains the same but data traffic between processes is reduced, resulting in a lower constant overhead. For a fixed number of processors and "sufficiently large" input, the $\frac{d}{k}\log\frac{d}{k}$ term dominates so the speedup is asymptotically linear in k.

On a shared-memory multiprocessor, the more efficient iterative algorithm can be parallelized in (looping) steps 2, 3, and 4, giving a running time $\Theta(\log d)$ for $\Theta(d)$ processors.

Shared memory is not absolutely necessary for an efficient parallel *FFT*. The iterative algorithm only requires certain combinations of values at each step; specific permutation networks with nearest-neighbor shared memory are sufficient. An example of such a processor is the BBN Butterfly [4].

Kung [12] suggested using special-purpose VLSI hardware for computing the *FFT* and other functions. Such circuits [2] can be very fast, but are only good for fixed-sized input and finite computation structures with bounded modulus. Such hardware would be useful for raising the bottom level of the *FFT* recursion from size 1 to the size s handled by the *FFT* hardware, with a resulting time complexity of $\Theta(n\log\frac{n}{s} + \frac{n}{s}\log s)$ for $\Theta(s\log s)$ hardware. It is unlikely that special hardware would be cost-effective, and would certainly be used only with great inconvenience.

2 Parallel Algorithms for Gröbner-Basis Reduction

The process of Gröbner-basis reduction converts a set of polynomials into a canonical basis set which generates the same ideal. It is a powerful tool for solving problems in connection with systems of multivariate polynomials [3], such as solving systems of nonlinear equations with arbitrarily-many variables and arbitrary degree, and simplification of polynomial expressions subject to polynomial equality side-relations.

The problem of generating a reduced Gröbner basis is known to be hard in exponential space; one lower bound on the cardinality of the reduced basis is doubly-exponential in the cardinality of the original basis. No upper bound is known on the cost of Gröbner-basis reduction; the best known algorithm (shown here) has been proven to converge without providing a bound on how long convergence takes:

Input: A set of polynomials F over R, an ordering $<$
Output: G, a set of polynomials representing the reduced basis of the ideal generated by F.

```
G:=F;
B:= {(f1, f2)|f1, f2 ∈ G, f1 ≠ f2}
While B ≠ φ do
        let (f1, f2) := a pair in B;
        B := B − {(f1, f2)};
        h := SPolynomial(f1,f2,<);
        h' := NormalForm(G,h);
        if h' ≠ 0 then
            B := B ∪ {(g, h')|g ∈ G};
            G := G ∪ {h'};
```

Figure 1 – Buchberger's algorithm for performing Gröbner-basis reduction

The *S-polynomial* of polynomials p and q is $ap - bq$, where the leading monomials (in some admissible ordering [7]) of ap and bq are (identically) the LCM of p and q. The *NormalForm* of p w.r.t. a basis set is the irreducible form of p, such that no basis element r and monomial c exist such that $p - cr = q$ and $q < p$ in the ordering.

2.1 Parallel Forms of Buchberger's Algorithm

Since the complexity of the Gröbner-basis problem is not understood, we cannot pursue any "optimal" parallel algorithms; instead we may look for parallel algorithms which outperform the best known serial ones. Watt [16] proposed a direct parallelization of Buchberger's iterative algorithm, effectively computing all S-polynomials in parallel, and reducing all basis elements w.r.t. each other in parallel:

```
G' := φ;
G := F;
while G ≠ G' do
    G' := G;
    B := {(f1, f2)| f1, f2 ∈ G, f1 ≠ f2}
    H := G;
    forall (f1, f2) ∈ B do
            h := SPolynomial(f1,f2);
            h' := NormalForm(G,h);
            if h' ≠ 0 then H := H ∪ {h'};
    G := φ;
    forall h ∈ H do
            h' := NormalForm(H − {h},h);
            if h' ≠ 0 then G := G ∪ {h'};
return G;
```

Figure 2 – Watt's parallel algorithm

Unfortunately this algorithm is incorrect. It discards the original basis after generating all pairwise S-polynomials; although the S-polynomials are elements of the original ideal, they may not generate the full ideal of the original basis set. A simple example is

$$\left\{ \begin{array}{c} x+y \\ x+y \end{array} \right\}$$

in which case the S-polynomials are all 0, yielding an empty basis. The correct reduced Gröbner-basis is $\{x+y\}$.

Two corrected algorithms are stated as follows:

```
G' := φ;
Repeat until G = G'
        G' := G;
        Simultaneously generate all S-polynomials from basis G,
                        and reduce the result w.r.t. G;
        Insert each S-polynomial into G;
        Fully reduce G;
Return G;
```

(which we will refer to as *Parallel S-polys*) and

Repeat until G no longer can change
  Choose two polynomials f & g from G;
  Take their S-polynomial h;
  Reduce h w.r.t. G;
  Simultaneously reduce each element of G w.r.t. h;
  Insert h into G;
  Reduce G w.r.t. itself;
Return G

which we will refer to as *Parallel Reduction*. Both of these are inherently "less parallel" than Watt's algorithm since they serialize the process of keeping the basis set reduced. Worse, they delay convergence over the serial algorithm, potentially causing them to run much slower in spite of the parallel operations. In tables 2-4 we compare their performance to a good serial implementation in Macsyma due to Gail Zacaharias [18].

The Zacharias implementation keeps the basis elements ordered, always reduces using the "smallest" polynomials first, and forms S-polynomials of the "smaller" basis elements first. In practice this appears to generate the key polynomials more quickly, which reduce the remaining basis elements, causing the basis to contract in degree and cardinality. The "parallel" algorithms defer this contraction by generating too many S-polynomials at once (since many of them may be eliminated by results of earlier S-polynomial operations) or attempting reductions by "larger" polynomials when "smaller" ones would otherwise be generated and used.

Table 2 – Comparative running times (in seconds)

Test case	Zacharias time	Parallel Reduction serial time	Parallel Reduction parallel time	Parallel S-Polys serial time	Parallel S-Polys parallel time
1	0.316	0.600	0.600	0.416	0.316
2	0.416	1.083	1.066	1.250	0.933
3	0.216	0.383	0.366	0.383	0.333
4	1.716	1.316	1.233	0.833	0.766
5	27.266	125.533	119.616	171.750	108.250
6	381.950	1102.650	1040.467	1376.117	1001.434
7	11.350	34.850	32.817	36.867	36.867
8	31.416	569.983	569.033	541.300	523.050
9	9.416	416.450	277.817	342.516	337.033
10	69.417	727.817	714.700	147.550	133.717
11	173.234	641.867	641.334	547.917	386.267
12	3473.466	*	*	*	*

Garbage-collection time has been deducted from the execution.
In case 12 the two parallel schemes exhausted memory space before finishing.

Table 3 – Estimated parallelism in Parallel S-poly and Reduction algorithms

Test case	Parallel Reduction maximum	Parallel Reduction average	Parallel S-Polys maximum	Parallel S-Polys average
1	3	1.00	6	1.32
2	4	1.02	3	1.34
3	4	1.05	3	1.15
4	5	1.07	6	1.09
5	4	1.05	3	1.59
6	4	1.06	3	1.37
7	3	1.06	1	1.00
8	3	1.00	1	1.03
9	4	1.50	3	1.02
10	5	1.02	3	1.10
11	6	1.00	15	1.42

Table 4 – Other execution statistics

Test case	Zacharias S-Polys	Zacharias Reductions	Parallel Reduction S-Polys	Parallel Reduction Reductions	Parallel S-Polys S-Polys	Parallel S-Polys Reductions	# of terms in result
1	3	10	10	35	9	33	12
2	1	24	4	65	6	110	23
3	1	25	5	50	6	54	12
4	14	225	12	92	9	82	15
5	2	188	4	187	6	362	210
6	2	189	4	189	6	365	918
7	2	108	1	43	1	43	7
8	1	14	2	23	2	23	130
9	10	330	5	215	6	286	26
10	9	100	8	90	6	92	131
11	14	159	25	251	28	303	110
12	10	250	?	?	?	?	*

* Case #12 generated too many terms to deal with, well in the thousands.

2.2 Reduction Under Alternative Orderings

Another way to utilize parallelism would be to reduce the initial basis under different orderings to see which one is produced the quickest; the execution time of the algorithm is highly sensitive to the ordering used, as is shown in table 5.

However, getting the fastest reduced basis is not sufficient. If we asked whether two basis sets generate the same ideal, the answer is *yes* if they have the same reduced Gröbner basis under a *given* ordering. We may find that the fastest reduced basis for each comes out with a different ordering. A remaining possibility is to re-reduce under a common ordering, shown in table 6.

We can see that this indirect approach *can* be faster as well as slower, but this approach of using alternative orderings appears appears weak unless we can better understand the relation between input, ordering, and running time.

Test	Time	Max/Min Time	Ordering
1	0.316	1.26	$x > y > z > t > s$
	0.233		$x > z > y > t > s$
	0.300		$y > x > z > t > s$
	0.250		$y > z > x > t > s$
	0.250		$z > x > y > t > s$
	0.250		$z > y > x > t > s$
2	0.416	4.09	$x > y > z$
	0.416		$x > z > y$
	1.700		$y > x > z$
	0.583		$y > z > x$
	0.849		$z > x > y$
	0.499		$z > y > x$
3	0.216	4.86	$x > y > z$
	0.182		$x > z > y$
	0.550		$y > x > z$
	0.866		$y > z > x$
	0.482		$z > x > y$
	1.049		$z > y > x$
4	1.716	1.02	$x > y$
	1.749		$y > x$
5	27.266	16.00	$x > y > z$
	29.033		$x > z > y$
	27.900		$y > x > z$
	27.399		$y > z > x$
	436.250		$z > x > y$
	424.983		$z > y > x$
7	11.350	8.41	$x > y$
	1.349		$y > x$
8	31.416	1.05	$x > y$
	33.049		$y > x$
9	9.416	1.13	$x > y > z$
	9.817		$x > z > y$
	10.450		$y > x > z$
	10.617		$y > z > x$
	10.133		$z > x > y$
	10.183		$z > y > x$
11	173.234	66.22	$x > y > z$
	18.450		$x > z > y$
	72.434		$y > x > z$
	7.967		$y > z > x$
	2.616		$z > x > y$
	25.434		$z > y > x$

Table 5 – Times (in Seconds) to Reduce Under Different Lexicographic Orderings

Table 6 – Time (in Seconds) to Reduce Directly vs. Indirectly		
	Case 7	Case 11
Direct Reduction		
Ordering:	$x > y$	$x > y > z$
Time:	11.350	173.234
Indirect Reduction		
First Ordering:	$y > x$	$z > x > y$
Second Ordering:	$x > y$	$x > y > z$
Time:	1.283	2668.434

2.3 Conclusions

We can draw the following conclusions at this point:

- "Obvious" ways of parallelizing Gröbner-basis reduction, such as generating S-polynomials or performing reductions in parallel sweeps are ineffective. Reducing under different orderings has some limited potential.

- Algorithmic "hacks" are effective in speeding up the process to a small degree; these give better performance than parallelism does, and seem to be themselves difficult to parallelize.

There is still room for exploring other ways to parallelize Gröbner-basis reduction, such as parallelizing the individual polynomial operations; we have explored the obvious ways and they are not obviously effective. It is possible that serial algorithms as or more efficient than Buchberger's will be devised, possibly being more parallelizable.

References

[1] Aho, A.V., Hopcroft, J.E., and Ullman, J.D. (1974). The Design and Analysis of Computer Algorithms. Addison-Wesley, Reading, MA.

[2] Bonneau, R.J. (1974). Polynomial Operations using the Fast Fourier Transform. Ph.D. thesis, Dept. of Mathematics, Mass. Inst. of Tech., Cambridge, MA.

[3] Buchberger, B. (1985). Gröbner Bases: An Algorithmic Method in Polynomial Ideal Theory. In Progress, Directions and Open Problems in Multidimensional Systems Theory (Ed. N.K. Bose), 184-232. D. Reidel Publishing Co., Hingam, MA.

[4] Butterfly Parallel Processing Computer. BBN Labs, Cambridge, MA.

[5] Char, B.W. et al. (1984). On the Design and Performance of the Maple System. Dept. of Computer Science, University of Waterloo, Waterloo, Ontario, Canada, CS-84-13.

[6] Cole, R. (1986), Parallel Merge Sort. In IEEE FOCS '86, 511-516.

[7] Dube, T., Mishra, B., and Yap, C.K. (1986). Admissible Orderings and Bounds for Grobner Basis Normal Form Algorithms. Tech. report #258, NYU Dept. of Computer Science.

[8] Fateman, R.J. (1974). On the Computation of Powers of Sparse Polynomials. Studies in Applied Mathematics LIII/2, 145-155.

[9] Fateman, R.J. (1974). Polynomial Multiplication, Powers and Asymptotic Analysis: Some Comments. SIAM J. Comput. 3/3, 196-213.

[10] Goto, E. and Kanada, Y. (1976). Hashing Lemmas on Time Complexities with Applications to Formula Manipulation. Proc. SYMSAC '76, 154-158. ACM, New York.

[11] Harper, L.H., Payne, T.H., Savage, J.E., and Strauss, E. (1975). Sorting X+Y. CACM 18/6, 347-349.

[12] Kung, H.T. (1981). Use of VLSI in Algebraic Computation: Some Suggestions. Proc. SYMSAC '81, 218-222.

[13] Moenck, R.T. (1976). Practical Fast Polynomial Multiplication. Proc. SYMSAC '76, 136-148.

[14] Ponder, C. (1988). Evaluation of "Performance Enhancements" in Algebraic Manipulation Systems. (Ph.D. Thesis) Report #88/438 UC Berkeley Computer Science Division.

[15] Vuillemin, J. *private communication* 7/31/87.

[16] Watt, S.M. (1986). Bounded Parallelism in Computer Algebra. Research report CS-86-12, Univ. of Waterloo Faculty of Mathematics.

[17] Winograd, S. (1978). On Computing the Fast Fourier Transform. Math. Comput. 32/141, 175-199.

[18] Zacharias, G. (1978). Generalized Gröbner Bases in Commutative Polynomial Rings. Bachelor thesis, Lab. for Computer Science, MIT, Cambridge MA.

Implementation of the LISP-Arbitrary Precision Arithmetic for a Vector Processor

Winfried Neun Herbert Melenk

Konrad-Zuse-Zentrum für Informationstechnik Berlin
Heilbronner Str. 10
D-1000 Berlin 31
Federal Republic of Germany

Abstract

Portable Standard LISP (PSL, Version 3.4) and REDUCE 3 were implemented for CRAY 1 and CRAY X-MP computers at the Konrad-Zuse-Zentrum für Informationstechnik Berlin in 1986. As an special aspect of the implementation of PSL, an interface to the vector hardware of CRAY processors was defined. With that interface and mostly driven by the needs of REDUCE applications (e.g. extensive calculations of Gröbner bases), the arbitrary precision integer arithmetic of PSL was rebuild using full power of the vector hardware. A modular arithmetic using vector hardware was also constructed.

1. Introduction

The Version 3.4 of Portable Standard LISP [1], a dialect of LISP developed at the University of Utah, was implemented for CRAY-1 and CRAY X-MP computers at the Konrad-Zuse-Zentrum für Informationstechnik Berlin (ZIB) in 1986 for the operating system COS and in 1987 for the operating system UNICOS. This implementation was a continuation of the earlier work on PSL version 3.2 at the University of Utah, Los Alamos Laboratories and CRAY Research Inc.

COMPUTER ALGEBRA AND PARALLELISM ISBN 0-12-221100-6

[2]. PSL 3.4 also serves as a supporting LISP system for REDUCE, which is distributed for CRAY computers by ZIB Berlin [3].

The CRAY version of REDUCE 3 is one of the fastest implementations available measured by the Reduce Standard test [4]. This test does not require nonstandard integers at least on a CRAY computer with its 64 bit words. With the new version, REDUCE 3.3 [5], the calculation of Gröbner bases was included and became a focus of the "Symbolic" group activities at ZIB. Some runtime analysis pointed out that the dominant part of the cpu consumption was caused by the integer arithmetic with arbitary precision, the PSL "bignums". One approach to solve "bigger" problems with Gröbner bases was to speed-up the arithmetic. It was not too difficult to get the idea of using the power of CRAY vector hardware for bignum arithmetic.

Some runtime analysis found out that the bignums which occur during Gröbner bases calculations are of moderate sizes which means their representation has some hundred decimal digits, in most cases below 1000 digits. Our task is different from those problems that occur when the calculation of PI to millions of digits is attempted, i.e. the asymptotic behaviour is not our main interest.

Since we cannot expect that the CRAY hardware architecture is known, we give a very short introduction, which will give some points for the later discussion on the arithmetic.

2. CRAY Hardware Outline

CRAY-1 processors [6] were first delivered in 1976 and are characterized and distinguished from usual computer systems by the main features: *functional units*, *instruction pipeline* and *Vector registers.* The following description is a very rough summary based on the CRAY X-MP architecture, without any idea of being 100% exact or complete.

Functional Units. The "classic" cpu is divided into several specialized units (in this case 13) which are able to do one special task each. A unit can operate independently from other units in parallel, e.g. the add integer can work in parallel with the floating point add, given the fact that the destinations and sources are disjoint. There are no functional units for multiplication integer or division integer, these have to be done via float operations.

Instruction Pipeline. The computer is (potentially) able to issue one instruction each cycle (or each two cycles), which is about 100 millions per second. This is true if no jumps are taken out from the sequence of instructions and if the functional units are not blocked by earlier instructions. To sustain this instruction rate, instructions have to be buffered in cpu and a jump to a location outside the instruction buffers will cause a tremendous delay.

Vector Registers. The CRAY computer has 8 vector registers which contain 64 (64 bit) words and which are operated by functional units with a rate of one per cpu cycle giving a dramatic speed-up result for some computations in FORTRAN programs. The result of a vector operation may be used directly by another vector operation before the first instruction is finished (chaining).

Note on the instruction timings. The memory access is slow in comparison with the arithmetic operations, just to give some idea: loading an integer from memory takes 13 cycles, add integer takes 3 cycles.

Examples:

a) The speed-up factor between scalar and vectorized quotient of two vectors is 5 for a modest vector length of 10 items.

b) Copying a vector from location a to location b takes about the same time as adding two vectors $a + c$ and storing the result to b, which is also the approximate time to be consumed if we multiplied a and c and stored it to b.

c) In many cases the estimation of the load time is a good estimation for the overall cpu-time and since many operations may be done in the shadow of a load.

Some impacts (programming rules) which were realized in parts automatically in CRAY PSL 3.4 via a scheduling path in the PSL compiler:

- Avoid access to memory whenever possible.
- Try to write the LISP operations so that they intermix the usage of funtional units and registers.
- Avoid jumps whenever possible.

On system programming level with vectors:

- Try to keep data in registers especially in vector registers without storing in memory.

A general idea is that an analysis of an algorithm with respect to its number of operations is not the whole story on a vector processor (see example b above). However, the simplicity of the algorithm and its data structure may be more important. It is well-known that a more sophisticated algorithm which saves operations is often hard to recode with a vector computer as a target hardware.

3. PSL Bignums

Portable Standard LISP's bignum package was developed by M. L. Griss, B. Morrison and A. C. Norman in 1982. It uses representation for its arbitrary precision integers which is very well suited for CRAY vector operations: a PSL vector (which is a number of LISP items stored continuously in heap memory) with a number of bits in each item, such that the product of two items can be computed without loss of bits into one word :

[length — sign — small integer — small integer— ... — small integer]

Convention: all small integers are positive.

With CRAY, the bound for the small integers is 2**21, so 43 bits in a word are lost. This is due to some curiosities in the arithmetic hardware. The multiplication of integers has to be done via the floating point format since there is no functional unit for multiplication integer or division integer. The floating point format on CRAY computers is 16 bits for sign of mantissa and exponent and 48 bits for mantissa. The reason why fast multiplication and division integers are possible with at most 46 bit results which restricts the word size for the bignums. A compression and decompression "on the fly" seems to be too expensive.

3.1 Example for system programming with vectorized bignums

As a first simple example, let us assume that we have to add two positive bignums with the same length – less than 63. The notation is described in the Usage of Vector [7] by H. Melenk and W. Neun.

We load the integers into two different vector registers and add the registers:

```
(VsetVL actuallength)                 % length of both numbers
(Vload (vreg 1) arg1 1)               % load numbers from arg1
                                        into V1
(Vload (Vreg 2) arg2 1)               % same with arg2 and V2
(VplusV (Vreg 0) (Vreg 1) (Vreg 2))   % V0 = v1 + v2
```

Now we have to worry about the carries:

```
(Vrshift (Vreg 3) (Vreg 0) 21)        % shift all items of a Vector
                                      % register by 21 into another one
(VandS (Vreg 4) mask (Vreg 0))        % mask out overflow bits
                                      % mask = 2**21 -1
```

Up to this point all operations were done in a chaining mode, i.e. the last item is read after the first has left the vector logic functional unit. We have now the following register contents:

V4 is the vector of sums cleared from carries.

V3 is the vector of carries which must be shifted by 64 bits into the "right" neighbor word, so that the addcarry does a true "carrying".

If the highmost digit of the carry is not zero, we have to increase size of operation by 1. This code is not presented here explicitly.

```
      (setq Vmask (VtestN (Vreg 3)))               % a mask indicating non-
                                                     zero carries
                                                   % if zero, we are ready
      ADDCARRY
      (while (not (eq Vmask 0)))                   % until all carries eaten up
      (when (overflown?) (increase-length))
   *  (Vdrshift (Vreg 5) (Vreg 3) 64)
      (VplusV (Vreg 6) (Vreg 4) (Vreg 5)           % the addcarry
   *  (Vrshift (Vreg 3) (Vreg 6) 21)               % shift carries to the right
  **  (setq Vmask (VtestN (Vreg 3)))               % a mask indicating
                                                   % nonzero carries
  **  (VandS (Vreg 4) mask (Vreg 0))               % mask out overflow bits
      (Vstore (Vreg 4) result 1)                   % finally get rid of result
```

In the add-carry loop, the instructions indicated by * and ** respectively use the same functional unit in order for the second operation to wait, but multiple carry propagation seldom occurs.

When the operand's length exceeds 63 items we subdivide the problem into pieces of 63 items, where each problem delivers the carry to the next subproblem. From this simple example we can see that the vector coding produces lots of instructions and the code is not very good readable. Fortunately, it is fast: the ratio between scalar and vector addition of bignums is shown in the figure.

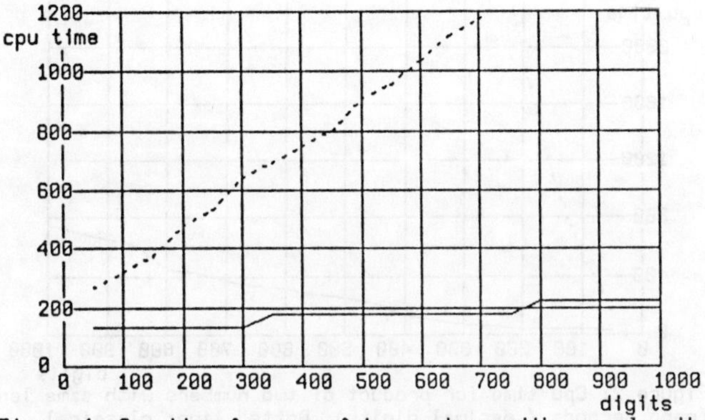

Figure 1: Cpu time for sum of two numbers with same length (nano seconds / decimal digits). Dotted line: classical version, full line: vectorized version

The unsmooth parts of the curve are caused by the 64 words length of the vector registers.

3.2 Basic algorithms for bignum arithmetic

The algorithms for multiplication of bignums have been discussed by many authors, e.g. [8]. In spite of the common recommendation to use the Fourier transform approach, we implemented for PSL on CRAY the "vectorized schoolboy" method. This was done because this method is simple and therefore "easily" vectorizable. Likewise, the lengths of the bignums which are used by our REDUCE applications are not too great, that the asymptotic behaviour of the algorithms are not of interest in almost any cases. In multiplication we can get benefits from the modest bits per item size we use because we can add up the partial products "infinitely" without an overflow. There is a maximum of 64 partial products that form the final result, so we can do the carry propagation when the final result is written to memory.

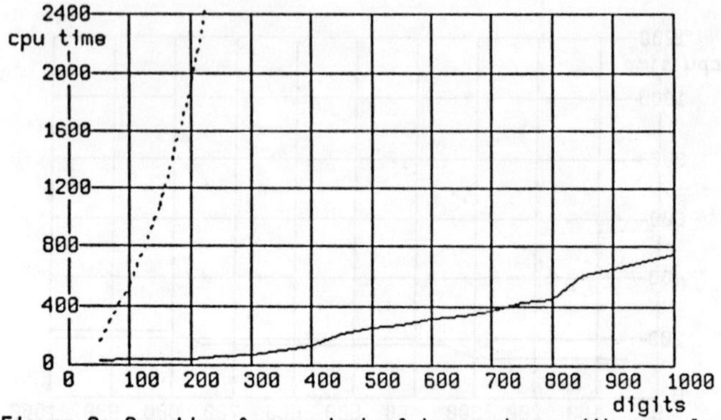

Figure 2: Cpu time for product of two numbers with same length (nano seconds / decimal digits). Dotted line: classical version, full line: vectorized version

Figure 2 shows the cpu consumption of the scalar and the vectorized algorithm. It can be seen that the $O(n^2)$ curve of the scalar algorithm is flattened dramatically so that the break even point between Fourier and RschoolboyS approach is far beyond the range of numbers in which we are interested in.

For quotient/remainder computation we use a two-fold approach: if the operands can both be held in vector registers, we use an algorithm that is based on the estimation of the result from the float quotient of the highmost digits of the bignums and keep the operands in vector registers all the time. In case the operands are bigger, we use a Newton iteration for the estimation of the reciprocal (shifted accordingly). This iteration naively coded had an interesting effect: the cpu time was about 4 times faster than Knuth's algorithm D [9], which is PSL bignum standard. But, in an overall summary it was a little bit slower since it used lots of intermediate results and furthermore, lots of extra garbage collections had been done. We solve that problem by explicite assignment of working memory.

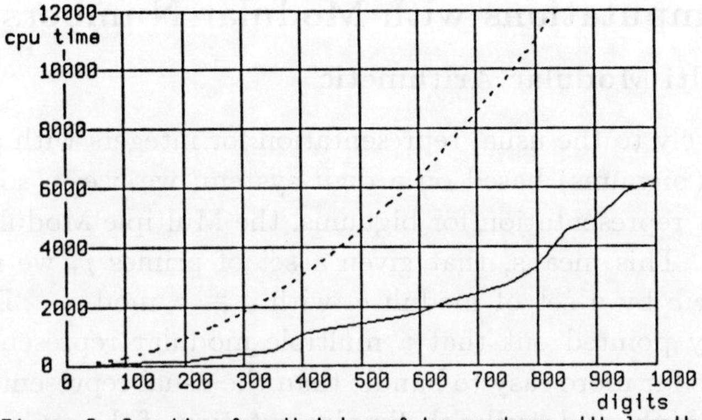

Figure 3: Cpu time for division of two numbers with length n and n/2 (nano seconds / n=decimal digits). Dotted line: classical version, full line: vectorized version

Figure 3 shows the behaviour of the vectorized algorithm – Newton approach vs. Knuth's D algorithm.

The gcd is used very often in rational arithmetic and causes heavy usage of the remainder algorithm. The gcd is computed via Euclid's algorithm. This is done mostly because it is a simple extension of the quotient/remainder algorithm described above. This means that, for operands less than 64 words the resulting gcd is computed completely in a vector register without using extra working memory. The binary algorithm (e.g. [9]) may be valuable, but our experiences with that algorithm have shown results that there are orders of magnitude slower than Euclid's algorithm even if we used vectorized add and shift operations. This may be caused by the fact that the shift operation of a bignum vector is not so easy and the "even" test which is required is expensive because of the access time to a single word in a vector register.

4. Computations with Modular Numbers

4.1 Multi Modular Arithmetic

Alternatively to the usual representation for integers with arbitrary precision (bignums) based on a digit system, we use in some cases a different representation for bignums, the Multiple Modular representation. This means, that given a set of primes p_i we represent an integer c by a set of moduli c_i with $c \equiv c_i \bmod p_i$. D. Knuth [9] already pointed out that a multiple modular representation of bignums is far more easy to handle than the usual representation. In this paragraph we describe the implementation of the multi modular arithmetic for REDUCE on a CRAY X-MP processor.

The arithmetic with these sets of moduli is defined elementwise

$$(a_1, \ldots, a_n)\#(b_1, \ldots, b_n) = (a_1\#b_1, \ldots, a_n\#b_n)$$
$$\text{where } \# \in \{+_{mod}, -_{mod}, *_{mod}\}\#$$

The multiple modular representation will return the same results as if we were doing single modular computations modulo the product of all primes p_i. The integer arithmetic handling multiple modular represented integers has a great advantage against the single modular computation with a big prime on CRAY systems: Arithmetic can fully exploit the vector hardware, if the primes are sufficiently small to be handled by hardware directly.

CRAY systems are equipped with vector hardware allowing up to 64 items containing up to 64 bits of information to be treated as one (vector) operand to e.g. an arithmetic operation, which then does the arithmetic on these items on a rate of about 100 Million operation/second. This way, using many primes instead of one very big one, we can use the computer hardware for computation directly. Therefore we choose the primes such that no overflow can happen when multiplying two moduli, which in our case restricts the primes to 21 bits.

Lets look at the multi modular addition as an example. Let V_i denote a vector (register) with up to 64 items. In REDUCE syntax the add operation is as follows (see [7]):

```
symbolic procedure multimodular!-add (src1,src2,primes,result);
begin integer nn;
   nn:= number of primes;
   while nn > 0 do
   ≪ VSETVL if nn>64 then 64 else nn; % set actual vectorlength
      Vload (V1,src1);
      Vload (V2,src2);               % load the operands
      VplusV (V3,V2,V1);             % do the add
      Vload (V4,primes);             % load the primes
      VminusV (V5,V3,V4);            % subtract primes from
                                       add result
      VsetVM VtestNEG V5;            % compute a mask
      VmergeVV(V0,V3,V5);            % merge the result
      Vstore (V0,result);
      nn := nn-64; increment(src1,src2,result) ≫;
end;
```

Comments. You will not be too surprised when you are told that this is an assembler program and most LISP functions are compiled into one instruction. We use the ability of the vector computer to intermix the operations. So almost all instructions can work in parallel except for the Vload that has to wait until the first two Vloads are ready. That means that we have a vectorized partially parallel operation.

Perhaps the only instructions we used that need explanation are the VsetVM and VmergeVV instructions. The VtestNeg produces a bit mask with bits set where V5's components are negative, those components where the add result is less than the prime component. The VmergeVV operation works under control of this mask and stores into V0 (componentwise) the item from V3 (the add result) or from V5 (the add minus prime result) where it is approppriate.

As the inverse of such a modular number we define (elementwise) the solution c_i given by the Extended Euclidian Algorithm (EEA) for:

$$c_i * m_i + x p_i = \gcd(m_i, p_i) = 1, \text{ since } p_i \text{ is prime.}$$

Therefore the computation of inverses is a mighty operation, computing the EEA of many pairs of numbers. The EEA computation can be done in vector mode too, in spite of the fact that not all the gcds will be ready at the same time. This is possible due to the fact that the vector hardware is able to work under the direction of logical masks indicating which elements are to be processed and which are not.

4.2 Computing Gröbner bases using Multiple Modular Arithmetic

The Gröbner base calculations with this type of arithmetic are done in following steps:

Primes must be chosen. Since the size of the primes is limited to 21 bits in the actual implementation, we have to get sufficiently many primes. If the final reconstruction points out that we have not enough primes, we have to redo all the computation. Unfortunately the sizes of intermediate polynomial coefficients are not predictable.

The Gröbner base calculation is done 'as usual' using functions that handle vectors of moduli instead of single numbers. During the calculation a situation may occur where almost all moduli are nonzero and some are zero or, the inverse problem, almost all moduli are zero and some are nonzero. This is caused by intermediate results, where the integer which is represented by the set of moduli is divided by one or more primes. That means that one or more primes are unlucky and have to be canceled. There is a need to decide now with which of the possible value to continue. We decide using a 'democratic principle' whether the number is zero or nonzero. The primes that are sentenced to be 'unlucky' are cancelled and not used in the final reconstruction phase. Since the number of primes used is normally large there is no real problem raised by that decision process.

The final result is a Gröbner base of polynomials with multiple modular coefficients which must be reconstructed afterwards. As described above we use the Chinese remainder algorithm to reconstruct the integer c from a set of moduli c_i with $c \equiv c_i \bmod p_i$ for all i, s.t. p_i is not cancelled. Wang's algorithm is started with the number c now to reconstruct the quotient. The algorithm detects an error if the quotient does not exist, i.e. the product of all primes p we used is not big enough to represent the coefficient of the Gröbner base solution. As described in the first step we have to redo the computation, but on the other hand we know the general layout of the Gröbner basis.

4.3 Examples

All examples have been computed by this trial and error method, but nevertheless the computer system time spent to get the result was small in comparison with the normal calculation. Unfortunately one cannot find many published examples of our special interest, i.e. with large bignums in intermediate polynomials, since most of these examples are beyond computability of conventional computer systems. We ran several examples, which because of the limited space cannot be shown here. All examples that run on the CRAY X-MP needed 60 to 400 primes (below 21 bits), i.e. the results were bounded by 240 to 2400 (decimal) digits.

We used as a test case the example 'Rose lexicographic' from [11]. It was started with 120 primes, 10 of which had to be canceled. The cpu time consumption was 2288 sec with normal arithmetic and 148 sec with multiple modular arithmetic plus 112 sec for reconstruction of the coefficients, giving a total of 260 seconds.

5. Conclusions

We presented our approach for arbitrary integer arithmetic on a special computer system. We feel that in the special environment of a vector computer some of the usual algorithm analysis do not give the correct results. Unfortunately, a correct analysis in our sense is to a large extent hardware dependent and will be obsolete with some new computer systems. Likewise, some cumbersome effects are due to our special hardware. On the other hand, a dramatic speed-up was achieved in some of our applications.

Recent improvements of the REDUCE factorizer done by J. Davenport include an interface for the computation with vectors of small integers. The implementation of that interface in CRAY vector code shows a remarkable speed-up in combination with Gröbner base calculations.

References

[1] M. L. Griss, A. C. Hearn: *A Portable LISP Compiler, Software Practice and Experience.* Vol. 11 (1981).

[2] J. W. Anderson, W. F. Galway, R. R. Kessler, H. Melenk, W. Neun: *Implementing and Optimizing LISP for the CRAY*, IEEE Software, July 1987.

[3] H. Melenk, W. Neun: *REDUCE User's Guide for CRAY 1/X-MP Series Running COS*, Konrad Zuse-Zentrum für Informationstechnik Berlin, Technical Report TR 87-4, August 1987.

[4] J. B. Marti, A. C. Hearn: *REDUCE as a LISP Benchmark*, ACM SIGSAM Bulletin, Vol. 19 No. 3.

[5] A. C. Hearn: *REDUCE User's Manual*, Version 3.3, The Rand Corporation (1987).

[6] *CRAY Computer Systems, X-MP Series Mainframe Reference Manual*, CRAY Research Inc., Mendota Heights (1983).

[7] H. Melenk, W. Neun: *Usage of Vector Hardware for LISP Processing*, Konrad Zuse-Zentrum für Informationstechnik Berlin (1986).

[8] J. D. Lipson: *Elements of Algebra and Algebraic Computing*, Benjamin/Cummings Publishing Company (1981).

[9] D. E. Knuth: *The Art of Computer Programming: Seminumerical Algorithms*, Vol. 2, 2nd Edition (1981).

[10] P. S. Wang: *A p-adic Algorithm for Univariate Partial Fractions*, ACM, Proc. SYMSAC (1981).

[11] W. Böge, R. Gebauer, H. Kredel: *Some Examples for Solving Systems of Algebraic Equations by Calculating Groebner Bases*, J. Symb. Comp. 1986, Vol.2, pp. 86 – 98.

Automatic Generation of Parallel Code for the Warp Computer

Trevor B. Tan

and

Paul S. Wang[1]

Department of Mathematical Sciences
Kent State University
Kent, Ohio 44242

Abstract

GENW2 is a parallel code generator written in Franz lisp that runs under the Vaxima symbolic computation system. Given high-level algorithm specifications and symbolic expressions in list representations, GENW2 outputs W2 code for the Warp systolic array computer. GENW2 supports three high-level parallel programming paradigms: pipelining, multitasking and array processing. Parallel algorithms specified in these paradigms are automatically mapped onto the Warp architecture freeing the user from much details of interprocess communication and synchronization. A code template can

[1]Work reported herein has been supported in part by the National Science Foundation under Grants CCR-8714836 and EET-8714628 and in part under a Cray Research Grant 440668.

COMPUTER ALGEBRA AND PARALLELISM ISBN 0-12-221100-6

be specified by the user to render the output code in a designated format. Generated routines may involve declarations, I/O statements, flow control, data distribution, subroutines, functions and macros. Large expressions are automatically segmented to smaller pieces. An optional code optimization phase is available. The development of GENW2 is motivated by research in automatic derivation and generation of parallel code for finite element analysis. However the package can be used independently.

1. Introduction

Automatic code generation is an important technique for combining the powers of symbolic and numeric computing systems. Many scientific computations involve two stages. First, the mathematical formulation of the problem at hand produces procedures and formulas that define the particular problem. The procedures and formulas are then written into numeric code (for example FORTRAN) for fast execution. Recent research at Kent State University resulted in the GENTRAN system [6] that can generate f77, RATFOR [16] and C code from symbolic list representations. The GENTRAN system is available under both the VAXIMA [4] [11] and the REDUCE [5] [12].

While GENTRAN is very useful, it is not designed to generate code for the advanced parallel computers that are becoming increasingly important. Automatic generation of parallel code not only reduces manual mathematical manipulations but also helps engineers and scientists who are not computer experts take advantage of advanced parallel computers. Our motivation is the desire to create an automatic coding system to generate parallel programs for finite element computations on available parallel computers such as the Warp computer, a systolic array computer of high performance processors developed at Carnegie Mellon University (CMU).

GENW2 is now being used as a component of P-FINGER [13] [15] an experimental system to derive parallel code for finite element analysis. The input to P-FINGER consists of problem domain parameters which are used to derive a combination of sequential and parallel algorithm specifications and symbolic representations. Those parts of the solution procedure mapped to the Warp parallel computer are fed to GENW2 which outputs the corresponding programs in W2, a high-level language for the Warp computer. GENW2 is an independent package so it can be combined with other code derivation software for different applications.

2. The Warp Parallel Computer

The Warp computer we used at CMU consists of the Warp array, the interface unit and a general-purpose workstation host [10] The host is a Sun-3 integrated with the Warp array through a special high-speed interface unit. The Warp array contains ten identical high-performance processors connected in a systolic fashion. Each such processor is called a Warp *cell.* All cells execute in parallel. Each cell is a programmable processor capable of performing 10 MFLOPS. The cells are connected as a linear array. Each cell performs I/O only with adjacent cells. The first cell can obtain input form and the last cell can send output to the interface unit. Inter-cell I/O has high bandwidth (80 MB/sec). Each cell has a program memory of 8K instructions and a data memory of 32K words. The Warp array is controlled by the host computer through the interface unit which contains a large amount of memory for storing data and programs for the Warp array. The interface unit controls peripherals and transfers data to and from the Warp array with low operating system overhead. For example, the last cell can pass data through the interface unit to the first cell directly [1].

2.1. Warp Programming Environment

The Sun-3 workstation runs Berkeley Unix providing a modern multi-user programming environment [2]. The W2 language is used to program the Warp computer. An interactive command interpreter called the Warp shell can be used to access the Warp machine. In addition to usual operating system commands, the Warp shell provides commands to execute programs on the Warp array. The Warp shell supports implementors of the programming environment, developers of Warp programs and users writing applications that call Warp programs alike. A run-time system provides low-level support such as downloading object code to Warp and transferring data between the host and the Warp array. A debugger is also available which permits the user to set breakpoints and inspect variables at the source program level.

2.2. The Language W2

W2 [7] is a simple Pascal-like high-level language for programming the Warp array. W2 hides the low-level details of the Warp computer and provides a high-level abstraction for the Warp programmer. Using W2, a programmer can specify programs for each Warp cell and define inter-cell communication. It is the programmer's responsibility to devise an algorithm and map that algorithm to cell programs which can be executed in parallel efficiently. This is not a trivial task and is often central to finding a Warp solution to a problem. W2 is a convenient tool to program that solution.

The W2 compiler [8], written in common lisp, runs on the host Sun-3 workstation and produces object code which is downloaded into the Warp array through the interface unit controlled by the run-time system that forms part of the Warp programming environment. The execution of the Warp array is controlled by a host program (usually a C program) which runs on the Sun-3. The program interface between the host program and the W2 program is the W2 *module.*

A W2 module specifies a call to the Warp array. A module consists of a name, typed formal parameters, and a program for each Warp cell. Using the name and parameters the host program can specify a Warp array call defined by any W2 module and can pass data into and collect results from the Warp array. A module contains one or more *cell programs.* Each cell program specifies the program for a number of consecutive Warp cells. All cells can execute the same program (homogeneous) or different programs (heterogeneous) simultaneously. Aside from specifying the per-cell instructions, a cell program also specifies the inter-cell communication through the use of the *send* and *receive* primitives. A send statement outputs a 32-bit word to the next cell while a receive statement inputs a 32-bit word from the previous cell. There are two I/O channels **x** and **y** between each pair of neighboring cells. The first cell can use *receive* to obtain parameters passed by the host program and the last cell can use *send* to return results to the host program. An overview of the Warp system is shown in Figure 1.

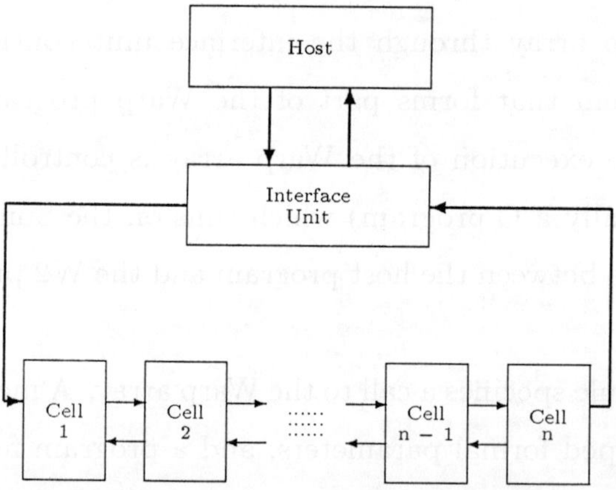

Figure 1: Overview of Warp System

3. Organization and Overview of GENW2

GENW2 is a code translation system written in Franz lisp. Running under the VAXIMA system GENW2 translates program specifications either in VAXIMA user-level or internal-level into equivalent W2 programs. The VAXIMA user-level specifications are, of course, processed by the VAXIMA evaluator and turned into simplified VAXIMA internal representation before reaching GENW2. Therefore, the front-end input to GENW2 is VAXIMA internal representation. GENW2 defines an intermediate code, called the GENW2 representation, as its own working language,. It is suitable for expressing algorithm specifications for the Warp computer and dependent on the syntax of W2. A program in VAXIMA internal representation is first converted into GENW2 representation by a simple transformation module, then generated into correct W2 code ready to compile and run on the Warp computer system. The GENW2 representation gives GENW2 a degree of system independence. To run GENW2 in a different symbolic computing system, simply make the input to GENW2 conform to the GENW2 representation by a system dependent translator. GENW2 has been used together with P-FINGER to generate W2 code for finite element analysis.

An overview of GENW2 is shown in Figure 2. For each phase, there is an independent program module. Phase O is small and the only VAXIMA dependent part of GENW2. All other GENW2 operations are based on the GENW2 representation and should run under most Franz lisp environments.

GENW2 also features template processing, code optimization, code segmentation as well as I/O redirection. These facilities make

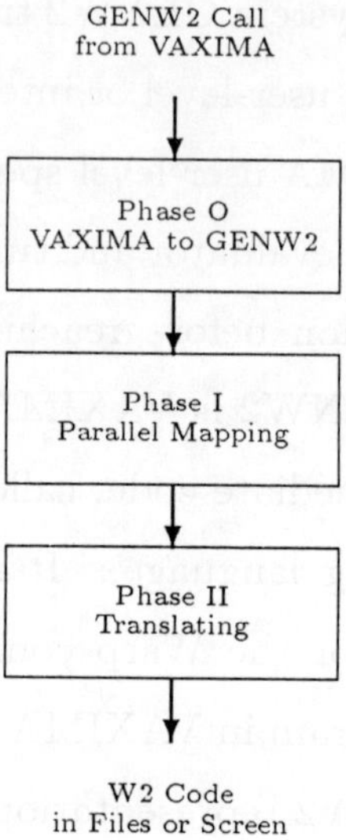

Figure 2: Overview of GENW2

it easy to obtain W2 code in user-specified formats and to deliver the target code to indicated files. GENW2 is invoked through one of several top-level functions or commands. Controlled by the programmer, each command can take input specified in VAXIMA user-level representation, VAXIMA internal representation or GENW2 representation.

The function of each phase will now be described. The input to phase O is in VAXIMA internal representation. This input is evaluated by VAXIMA and the result is transformed into GENW2 representation, which is passed to the next phase. This is a system dependent processing phase.

Several important functions are performed in Phase I. It implements syntax checking and three high-level parallel programming paradigms: *pipelining*, *multitasking* and *array processing*. First, the input is checked recursively for conformity to GENW2 representation syntax. A message is displayed if an error is found. Then any program specifications using the supported paradigms are processed. Basically, such program specifications are rewritten as appropriate lower-level specifications involving cellprograms and I/O between neighboring cells as well as the host. These high-level parallel paradigms provide a way to specify parallel programs without regard to the specific architecture of the Warp computer [9]. GENW2 takes care of supplying the detailed and sometimes difficult program mapping and I/O routing necessary to achieve the desired parallel effect. Detailed descriptions of these paradigms are contained in section 4. The output of this phase is still in GENW2 representation.

The main tasks performed in phase II are: preprocessing, code optimization, code segmentation, literal translation and postprocessing. After the first three tasks, the modified GENW2 representation is ready to be literally translated into W2 code. Each GENW2 representation is translated by its dedicated translating function and some formatting functions, while information obtained in the preprocessing phase is also used in constructing a complete W2 program. The generated W2 code is represented internally by a list structure with proper formatting specifications. The postprocessing delivers the generated W2 code into the specified output files. The result is syntactically correct and well-formatted W2 code, ready to be compiled and run on the Warp computer.

4. Mapping of Parallel Paradigms

A user can supply input to GENW2 that can be translated literally into W2 code. This gives the user maximum control over the code generated. To provide more convenience, GENW2 also supports three high-level parallel program paradigms: the pipelining mode, the multitasking mode, and the array processing mode. Input specified in any of these three modes are first expanded in phase I into a specification involving the necessary low-level primitives supported by W2. This can then be translated literally into W2.

4.1. Pipelining Mode

In the pipelining mode, the input specifies a number of ordered *steptasks*. Each steptask requires the result of its previous steptask as input. Input to the pipeline of steptasks is given to the first steptask. Output of the pipeline is produced by the last steptask. There are no other communications among the step tasks or the host. There can be many steptasks up to the maximum number of Warp cells.

A typical steptask and its syntax is shown here.

```
(steptask task_name)

(type var_type var_a var_b ... var_x var_y ...)

;; statements to be translated into part of a cellprogram

(input var_a var_b ... )
```

```
;; more statements to be translated into part of a cellprogram

(output var_x var_y ... )

;; more statements to be translated into part of a cellprogram

(endsteptask)
```

A group of steptasks are given in sequence sandwiched between the statements (pipe pipe_name) and (endpipe). The inter-task I/O variables are declared in each steptask and the number and the types of the corresponding I/O variables must match. An I/O variable, such as `var_a`, can be a scalar, a 1-D or a 2-D array. The input statement for the first steptask and the output statement for the last steptask are special:

`(hostinput cell_var host_var)` for first steptask, and
`(hostoutput cell_var host_var)` for the last steptask.

The high-level inter-task communications specifications are translated into correctly synchronized W2 *send* and *receive* statements in the generated cellprograms, making programming the Warp much easier.

4.2. Multitasking Mode

Another convenient parallel paradigm is multiple slave tasks controlled by a master process. The slave tasks are independent and execute in parallel. Each slave task can be viewed as a subroutine called by the control program. A slave task corresponds to a program for a single cell. The maximum number of slave tasks is limited to the number of Warp cells.

The syntax of a slave task is as follows.

```
(task task_name)

(input cell_var_a host_var_a)
(input cell_var_b host_var_b)
...

;; statements to be translated into W2

(output cell_var_x host_var_x)
(output cell_var_y host_var_y)
...

(endtask)
```

As before, an I/O variable, such as `cell_var_a`, is a scalar or an array of dimension 1 or 2. The types of cell variables are declared using the type statement as in the pipeline mode. The input, output and type statements can appear anywhere outside a function in the body of a task. The slave tasks are specified in any order between the pair of statements:

```
(mtask mtask_name)
(endmtask)
```

If the slave tasks run the same program on different data supplied by the control program, the SIMD model can be simulated. If the slave tasks run different programs, a MIMD model results. In generating the W2 programs for a mtask, GENW2 produces the correct data propagation mechanism to simulate direct I/O between the host and each individual task.

4.3. Array Processing Mode

In the array processing mode, the user can map certain usual operations on vectors (1-D array) and matrices (2-D array) on the Warp array. GENW2 will generate the appropriate W2 programs and the necessary communications between cells to take full advantage of the Warp architecture. Because of the overhead involved in calling Warp, no significant speed up can be gained for array sizes much smaller than 10,000. Of course this depends on the type of operation involved also. Some supported operations are listed here [3]. Other useful operations will be implemented in the future.

```
GENW2 (DOTPRODUCT (VECTOR_A, VECTOR_B, PRODUCT, DIM {,[PORTS]}));
GENW2 (MATADD (MATRIX_A, MATRIX_B, MATRIX_C, DIM_1, DIM_2 {,[PORTS]}));
GENW2 (MATSUB (MATRIX_A, MATRIX_B, MATRIX_C, DIM_1, DIM_2 {,[PORTS]}));
GENW2 (MATMUL (MATRIX_A, MATRIX_B, MATRIX_C, DIM_1, DIM_2 {,[PORTS]}));
GENW2 (ALL (ARRAY, FLAG, DIM_1, DIM_2 {,[PORTS]}));
GENW2 (ANY (ARRAY, FLAG, DIM_1, DIM_2 {,[PORTS]}));
GENW2 (SUM (ARRAY, SUM, DIM_1, DIM_2 {,[PORTS]}));
GENW2 (MAXVAL (ARRAY, VALUE, DIM_1, DIM_2 {,[PORTS]}));
GENW2 (MINVAL (ARRAY, VALUE, DIM_1, DIM_2 {,[PORTS]}));
```

where [PORTS] is an optional output file list. The default output port is the user's terminal. These GENW2 calls are designed to generate appropriate W2 modules for each of the following array or vector operations, respectively.

Dot product of vectors

Matrix addition and subtraction

Matrix multiplication

Test if all the array or vector elements are true

Test if any entry of the array or vector is true

The summation of all the array or vector elements

The maximum value of the array or vector elements

The minimum value of the array or vector elements

5. The Translation Process

Now let us turn our attention to the translation process used by GENW2. Figure 3 shows an overview of the translation process.

The translation process assumes that the input is already in GENW2 representation and that statements can be translated literally. This phase involves several steps each performs its transformation on the input code and passes the transformed code to the next step.

First, any indicated output files are opened, some symbol tables, variables and flags are set up. Code optimization comes next [17]. Common subexpressions in the input are identified and replaced by system generated variables. Assignment statements involving these temporary variables are inserted into the input. The following step is expression segmentation which breaks input expressions that are too long down to several subexpressions of manageable size. The optimized and segmented code is translated literally into W2 expressions and statements. The code module for literal translation contains many translating routines. For example there are routines for program headings, functions/macros, statements and expressions. The translating routines also call other formatting functions to construct W2 code. Certain statements such as type declarations are also inserted at this time. Finally, the generated W2 code is output.

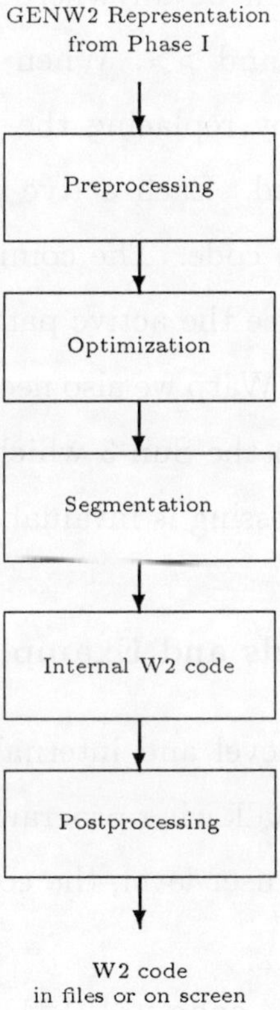

Figure 3: Overview of Translation Process

6. Template Processing

A template file is a device for a user to specify the exact format of the code produced as well as how automatically generated codes are combined with fixed code. A template file consists of *fixed* and *active* parts. In a template file, everything not enclosed in << and >> is fixed. An active part can be anywhere in a template file as long as it is given inside << and >>. When processing a template file, GENW2 transforms it by replacing the active parts while leaving the fixed parts untouched. Each active part contains one or more commands that generate code. The commands are carried out and the generated code replace the active part in place.

In generating code for Warp we also need to generate control codes that run under UNIX on the Sun-3 which is in turn invoked by f77 routines. Template processing is invaluable in such a situation.

7. GENW2 Commands and Examples

Both VAXIMA user-level and internal-level commands are provided to invoke GENW2 allowing programs at both levels to request W2 code generation. At user-level, the command is:

```
genw2 (code_1, ..., code_n {,[file_1, ..., file_m]});
```

which generates `code_1, ...  code_n` into W2. One or more codes can be given each represented in VAXIMA user-level syntax and may involve VAXIMA-defined computations. Each code can be an individual statement or a complete cell program. The optional [`file_1`, `...`, `file_m`] argument list specifies one or more output files to use. The default file is the user's terminal. At the lisp level, this command takes the form:

```
(genw2 '(code_1 ...  code_n) '(file_1 ...  file_m))
```

where `code_1 ...  code_n` is a sequence of one or more input codes in GENW2 representation.

In addition to the above commands, there are commands to process template files as well as open and close output ports.

6.1. A Complete Example

Let us illustrate how GENW2 works further by giving a simple example. We shall show the input, all important stages of code transformation, and the final W2 code generated. The input to GENW2 specifies a program to compute the partial sums of 25 inner products of vectors of length 100. The program is specified using the pipeline paradigm. The 50 vectors are passed to the first cell from a host array `hostary`. The results, 25 partial sums, are stored in the host array `hostsum`.

6.2. Input from VAXIMA User-level:

```
genw2(
pipe (maxdotprodt),       /* Find the partial sums of 25 dot products */
                          /* Input array contains 25 pairs of vectors, */
                          /* each of length 100 stored in 50 x 100 2-D */
                          /* array -- hostary. output is stored in the */
```

```
/* vector -- hostsum. */

type(float,hostary[50,100],hostsum[25]),
steptask (times), /* first steptask */
type(float, cellary[50,100]),
type(float,p),
multi() := (
  type(int, i, j),
  for i : 0 thru 49 step 2 do (
    for j : 0 thru 99 do
      p : cellary[i,j] * cellary[i+1,j],
    output(p)) ),
hostinput(cellary, hostary),
call(multi),
endsteptask(),

steptask(addition), /* second steptask */
type(float, sum, term),
add() := (
  type(int, i, j),
  for i : 0 thru 49 step 2 do (
    sum : 0.0,
    for j : 0 thru 99 do ( input(term), sum : sum + term),
    output(sum))),
call(add),
endsteptask(),

steptask(stepsum), /* third steptask */
type(float, cellsum, inner),
total() := (
  type(int, i),
  cellsum : 0.0,
  for i : 0 thru 24 do (
    input(inner),
    cellsum : cellsum + inner,
    hostoutput(cellsum,hostsum[i]) )),
call(total),
endsteptask(),
```

```
endpipe() );
```

6.3. Input Converted to VAXIMA Internal Representation:

```
((($pipe) $maxdotprodt)
(($type) $float (($hostary array) 50 100) (($hostsum array) 25))
(($steptask) $times)                      /* first step task */
(($type) $float (($cellary array) 50 100))
(($type) $float $p)
((mdefine) (($multi)) ((mprogn)
(($type) $int $i $j)
((mdo) $i 0 2 nil 49 nil ((mprogn)
((mdo) $j 0 nil nil 99 nil
((msetq) $p ((mtimes) (($cellary array) $i $j)
(($cellary array) ((mplus) $i 1) $j))))
(($output) $p)))))
(($hostinput) $cellary $hostary)
(($call) $multi)
(($endsteptask))
(($steptask) $addition)                   /* second step task */

...

(($steptask) $stepsum)                    /* last step task */
(($type) $float $cellsum $inner)
((mdefine) (($total)) ((mprogn)
(($type) $int $i)
((msetq) $cellsum 0.0)
((mdo) $i 0 nil nil 24 nil ((mprogn)
(($input) $inner)
((msetq) $cellsum ((mplus) $cellsum $inner))
(($hostoutput) $cellsum (($hostsum array) i))))))
(($call) $total)
(($endsteptask))
(($endpipe)))
```

6.4. Input Converted to GENW2 Representation:

```
((pipe maxdotprodt)
(type float (hostary 50 100) (hostsum 25))
(steptask times)                        /* first step task */
(type float (cellary 50 100))
(type float p)
(defun multi nil (prog nil
(type int i j)
(do ((i 0 (plus i 2))) ((greaterp i 49)) (prog nil
(do ((j 0 (plus j 1))) ((greaterp j 99))
(setq p (times (cellary i j) (cellary (plus i 1) j))))
(output p)))))
(hostinput cellary hostary)
(call multi)
(endsteptask)
(steptask addition)                     /* second step task */

...

(steptask stepsum)                      /* last step task */
(type float cellsum inner)
(defun total nil (prog nil
(type int i)
(setq cellsum 0.0)
(do ((i 0 (plus i 1))) ((greaterp i 24)) (prog nil
(input inner)
(setq cellsum (plus cellsum inner))
(hostoutput cellsum (hostsum i))))))
(call total)
(endsteptask)
(endpipe))
```

6.5. After Mapping of Parallel Paradigms:

```
((module maxdotprodt (in hostary) (out hostsum))
(type float (hostary 50 100) (hostsum 25))
(cellprogram times 0 0)                 /* first step task */
```

```
(type float (cellary 50 100))
(type float p)
(defun multi nil (prog nil
(type int i j)
(do ((i 0 (plus i 2))) ((greaterp i 49)) (prog nil
(do ((j 0 (plus j 1))) ((greaterp j 99))
(setq p (times (cellary i j) (cellary (plus i 1) j))))
(send r x p)))))
(defun input0 nil (prog nil
(type int i j)
(do ((i 0 (plus i 1))) ((greaterp i 49))
(do ((j 0 (plus j 1))) ((greaterp j 99))
(receive l x (cellary i j) (hostary i j))))))
(call input0 multi)
(endcellprogram)
(cellprogram addition 1 1)                  /* second step task */

...

(cellprogram stepsum 2 2)                   /* last step task */
(type float cellsum inner)
(defun total nil (prog nil
(type int i)
(setq cellsum 0.0)
(do ((i 0 (plus i 1))) ((greaterp i 24)) (prog nil
(receive l x inner)
(setq cellsum (plus cellsum inner))
(send r x cellsum (hostsum i))))))
(call total)
(endcellprogram)
(endmodule))
```

6.6. Generated W2 Code:

```
module maxdotprodt (hostary in, hostsum out)
float hostary[50][100],hostsum[25];

cellprogram (times : 0 : 0)                 /* first step task */
```

```
begin
  float cellary[50][100],p;

  function multi                          /* element-wide multiplication */
  begin
    int i,j;
    for i := 0 to 49 by 2 do
      {
        for j := 0 to 99 do
          {
          p := cellary[i][j] * cellary[i + 1][j];
          };
        send(r,x,p);                      /* output to next step */
      };
  end

  function input0
  begin
    int i,j;
    for i := 0 to 49 do
      {
      for j := 0 to 99 do
        {
        receive(l,x,cellary[i][j],hostary[i][j]);
        };                                /* input from the host */
      };
  end

  call input0;
  call multi;
end  /*  end of cellprogram  */


cellprogram (addition : 1 : 1)           /* second step task */
begin
  float sum,term;

  function add
```

```
 begin
   int i,j;
   for i := 0 to 49 by 2 do
     {
       sum := 0.0;
       for j := 0 to 99 do
         {
           receive(l,x,term);          /* input from previous step */
           sum := sum + term;
         };
       send(r,x,sum);                  /* output to next step */
     };
 end

 call add;
end  /*  end of cellprogram  */


cellprogram (stepsum : 2 : 2)          /* last step task */
begin
 float cellsum,inner;

 function total
 begin
   int i;
   cellsum := 0.0;
   for i := 0 to 24 do
     {
       receive(l,x,inner);             /* input from previous step */
       cellsum := cellsum + inner;
       send(r,x,cellsum,hostsum[i]);   /* output to the host */
     };
 end

 call total;
end  /*  end of cellprogram  */

/*  end of module  */
```

8. Conclusion

GENW2 is a first attempt to generate parallel code from quantities derived in a symbolic system. Because of the complexity involved in writing parallel code, we believe automatic parallel code generation in problem specific domains can bring the power of modern parallel computer to a much wider user community. GENW2 has benefited from the research experience of GENTRAN. On top of literal translation into parallel code, GENW2 also provides three useful parallel programming paradigms that allow the user to conveniently specify parallel algorithms and to rely on GENW2 to map these algorithms to the Warp architecture. Another important feature of GENW2 is its organization into the system dependent front-end and the system independent translation engine. This makes it much more portable to other symbolic manipulation systems.

Although GENW2 is motivated by developing an automatic coding system for finite element analysis, it can be used independently or with other problem-specific code derivation systems.

It is hoped that the research reported here will lead to a more in depth examination of the issues involved in the automatic generation of parallel code and shed light on ways to generate parallel code for different hardware architectures.

Acknowledgement

We would like to thank Professor H. T. Kung and Dr. Thomas Gross of CMU for their help and cooperation and for access to the

Warp machine. We would also like to thank N. Sharma for useful discussions and for testing early versions of GENW2.

References

1. Annaratone, M., Arnould, E., Gross, T., Kung, H. T., Lam, M. S., Menzilcioglu, O., and Webb, J. A., "The Warp Machine: Architecture, Implementation and Performance", IEEE Trans. on Computers, Dec. 1987, vol. C-36, no. 12, pp. 1523-1538.

2. Bruegge, B., *Warp Programming Environment: User Manual*, Department of Computer Science, Carnegie Mellon University, Pitts., PA, July 1987.

3. *Fortran*, X3J3 / S8.104, American National Standard for Information Systems Programming Language, June 1987.

4. Gates, B. L., *The Design and Implementation of An Automatic RATFOR Code Generator for VAXIMA*, Master's Thesis, Department of Mathematical Sciences, Kent State University, Kent, Ohio, 1984.

5. Gates, B. L., "A Numerical Code Generation Facility for Reduce", Proceedings, ACM SYMSAC'86, July 21-23, 1986, pp. 94-99.

6. Gates, B. L., "GENTRAN Design and Implementation", Department of Informatics, Twente University of Technology, The Netherlands, August, 1985.

7. Gross, T. and Lam, M., "A brief Description of W2", Department of Computer Science, Carnegie Mellon University, Pitts., PA, June 1987.

8. Gross, T. Lam, M. and Reinders, J., *Programming Warp in W2*, Department of Computer Science, Carnegie Mellon University, Pitts. and G. E. RSD Court Street Plant, Syracuse, N. Y., 1988.

9. Hamey, L. G. C. Webb, J. A. and Wu, I. C., "Low-Level Vision on Warp and the Apply Programming Model", The Robotics Institute, Carnegie Mellon University, Pitts., PA, July 1987.

10. Kung, H. T., "The CMU Warp Computer", Supercomputers, Edited by Matsen, F. A. and Tajima, T., pp. 237 - 246, 1986.

11. *MACSYMA Reference Manual*, Version Nine, The MATHLAB Group, Laboratory for Computer Science, M.I.T., Cambridge, Mass., 1977.

12. *Reduce User's Manual*, Version 3.0, Edited by Anthony C. Hearn, The Rand Corporation, Santa Monica, California. April 1983.

13. Sharma, N. and Wang, P. S., "Symbolic Derivation and Automatic Generation of Parallel Routines for Finite Element Analysis", to appear in Proceedings, International Symposium on Symbolic and Algebraic Computation (ISSAC-88), Roma, Italy, July 4-8, 1988.

14. van Hulzen, J. A., "Code Optimization of Multivariate Polynomial Schemes: A Pragmatic Approach", Proceedings, EUROCAL'83, Springer LNCS series No. 162, pp. 268-300, 1983.

15. Wang, P. S., "FINGER: A Symbolic System for Automatic Generation of Numerical Programs in Finite Element Analysis", Journal of Symbolic Computation, vol. 2, 1986, pp. 305-316, Academic Press.

16. Wang, P. S. and Gates B. L., "A LISP-Based RATFOR Code Generator", Proceedings, The Third MACSYMA Users' Conference, 1984, pp. 319-329.

Automatic Code Vectorization of Arithmetic Expressions by Bottom-Up Structure Recognition

V.V. Goldman and J.A. van Hulzen

Twente University, Department of Computer Science
P.O. Box 217, 7500 AE Enschede, The Netherlands

ABSTRACT

A bottom-up structure recognizer for vectorizing or parallelizing scalar code typically generated by computer algebra processing is proposed. A small example is worked out and some important issues are discussed.

1. Introduction

An important application of computer algebra systems is the generation of numerical results via automatic or semi-automatic program generation. In a system such as REDUCE [1], for instance, the FORT switch will cause results to be output in FORTAN syntax. A flexible general-purpose symbolic-numeric package (GENTRAN) has been recently reported [2]. The development of problem-specific FORTRAN code generators for use in conjunction with large numerical libraries such as NAG and IMSL is also in progress [3,4]. One of the advantages of a symbolic-numeric interface is the help it provides in reducing tediousness and human error of hand-coding, especially for large problems.

Attendant to automatic program generation is the problem of automatic source-code optimization. This is a crucial aspect because code generated from symbolic computations often tends to be made up of lengthy arithmetic expressions. As discussed by Hearn, and others [5,6], structure recognition is an important element of optimization in the present context. It is now widely recognized that the advent of vector and parallel machines has given the optimization problem a new dimension. This is not only due to the radical difference between parallel machines and the traditional von Neumann type, but also to the diversity of the parallel architectures themselves. Each new architecture usually demands a

COMPUTER ALGEBRA AND
PARALLELISM ISBN 0-12-221100-6

rethinking of optimization strategies. The problem is further compounded by all sorts of extraneous parameters such as inner workings of the source language compiler, problem size, and the inherent nature of the problem and of the algorithm. A good discussion of vectorization issues based on a full-machine model has been given by Arnold [7].

Given a particular architecture, parallelism can be induced at various levels and various stages of the software development cycle. At the design level, potential parallelism will be detected as early as possible and a suitable parallel algorithm will be designed. Parallelization here is problem-specific and, the state of the art being what it is, does not lend itself very well to automatization. At the bottom of the spectrum is the parallelization of programs in coded form. Most work at this level builds mainly on the work of Kuck and coworkers [8]. The techniques are not necessarily problem-specific and are well suited to automatic processing. This paper adresses itself to this level. So far most efforts here have concentrated on the parallelization or vectorization of DO loops. Computer algebraic processing, however, can hide possibly implicit loop structures and inherent mathematical symmetries. Therefore if one wants to optimize computer algebra output at this level some data restructuring is necessary.

In this paper we report on preliminary work on a bottom-up structure recognizer for straight line code of arithmetic expressions, and indicate its relevance to automatic vectorization of scalar computer algebra output. The methods we shall be using are equally applicable to multiprocessor machines. For illustrative purposes, however, we restrict ourselves here to vector machines with efficient vector gather-scatter operations [9]. The recognition method is derived from van Hulzen's code optimization scheme as based on the Breuer grow factor algorithm for detecting common subexpressions. In van Hulzen's work a block of assignment statements is mapped onto incidence matrices. The code optimization is effected by a series of transformations on the matrices designed to reduce the arithmetic complexity, the most important being the recognition and extraction of common subexpressions. This work has been described elsewhere [10] and we shall restrict ourselves here to a short description and to the use of simple illustrations. It is worth keeping in mind however, that the intention is to process code which has been produced after extensive symbolic computations and which, as opposed to hand written programs, typically contains long blocks of lengthy expressions.

Besides being useful for common subexpression search, the incidence matrices form an ideal data type for abstracting other types of structures by means of heuristic transformations. The construction of a prototype for finding common structures using such a data representation has been reported [11]. This latter work was designed to retrieve functional applications in order to optimize symbolic computations involving multiple differentiation operations, and to retrieve symmetries which may have become hidden after symbolic processing [12]. In the present context we shall mean by common structures the occurences, in scalar code, of vectorizable operations such as simple arithmetic operations , linked triads, and standard library vector functions (sqrt, sin, cos, etc.). The idea is to collect all instances of a common structure of a given type and bundle them into a vectorizable DO loop within data dependency constraints. As already mentioned above the approach is heuristic, which means that it involves a hopefully well chosen sequence of transformations on the reprensentation of the code. A particular choice will depend on many parameters, the most important being the particular architecture. The nature of the available optimizing compiler is also essential. We are aware of the fact that the data reorganization just described can result in considerable overhead. At the moment we are experimenting with various strategies, one of which will be illustrated below. The program is being written in REDUCE and is an experimental extension to the earlier mentioned Optimizer.

2. The Optimizer

Conventionally, arithmetic expressions are represented by an abstract syntax tree where nodes represent operators and vertices point to terminals or to subexpresssions [13]. On the other hand, the incidence matrix representation of a set of arithmetic expressions is a two-dimensional structure where rows represent expression or sub-expression references and columns represent identifier references such as variable or function names. A given sub-expression will be entered in one of two types of incidence matrices, one for sums or one for products, depending on the nature of the arithmetic operation at the top level of the expression. The two matrices are correlated by auxiliary predecessor-successsor information at the row level for every subexpression reference. The actual entries in the matrices are either multiplicative numerical coefficients for the sums matrix, or powers for the products matrix. A more formal definition of this representation has been given elsewhere [10]. This representation has been chosen because it facilitates the application of a large variety of heuristic transformations on the code. As a trivial illustration consider the following set of assignment statements:

```
A:=B*C+D*E
F:=G*H+J*K
R:=D*E+G*H
```

The incidence matrix representation for this block is shown below:

		B	C	D	E	G	H	J	K	Predecessor	Successors
Sums	A									-	1,2
	F									-	3,4
	R									-	5,6
Products	1	1	1							A	-
	2			1	1					A	-
	3					1	1			F	-
	4							1	1	F	-
	5			1	1					R	-
	6					1	1			R	-

The Optimizer will, among other things, recognize common subexpressions (indicated within the boxes in the incidence matrices) and produce the following matrix representation:

		G1	G2	B	C	D	E	G	H	J	K	Predecessor	Successor
Sums	A		1									-	1
	F	1										-	2
	R	1	1									-	-
Products	1			1	1							A	-
	2									1	1	F	-
	G1							1	1			-	-
	G2					1	1					-	-

which is equivalent to the following code.

```
G2:=D*E
A:=G2+B*C
G1:=G*H
F:=G1+J*K
R:=G1+G2
```

Consider now the code:

```
A:=B*C+D*E
F:=G*H+J*K
```

The matrix representation is similar to the previous example.

		B	C	D	E	G	H	J	K	Predecessor	Successors
Sums	A									–	1,2
	F									–	3,4
Products	1	1	1							A	–
	2			1	1					A	–
	3					1	1			F	–
	4							1	1	F	–

This representation can also be used to detect common structures. This is done by first imposing a partial ordering relation on the matrix entries for each row separately and then creating auxiliary matrices after some suitable topological sorting. In the present simple example the *auxiliary* matrices reduce to:

		X1	X2	Predecessor	Successors
Sums	A			–	1,2
	F			–	3,4
Products	1	1	1	A	–
	2	1	1	A	–
	3	1	1	F	–
	4	1	1	F	–

Common structures are now detected and processed in the same way common subexpressions were in the previous example. As can be seen by the representation above, products are detected first. The next iteration will reveal the sums. This finally leads to the following code:

```
F0(X1,X2)=X1*X2
F1(X1,X2,X3,X4)=F0(X1,X2)+F0(X3,X4)
A=F1(B,C,D,E)
F=F1(G,H,J,K)
```

The fact that a partial ordering is involved implies that much freedom exists and the processing must rely on heuristics. For large problems the present implementation [11] works best when preceded by common subexpression extraction. Further work is in progress.

3. Vectorization

The relevance to vectorization or parallelization is further illustrated below by means of a small example which consists of computing a rotation matrix resulting from four successive rotations about various axes. We show how a small number of simple transformations vectorizes the code. Some of the operations have been discussed in the literature (see for instance [14]). In this special example we restrict ourselves to rotations about the y and z axes. The rotation matrices, as a function of the rotation angle α, are respectively:

$$R_y(\alpha)=\begin{bmatrix} cos(\alpha) & 0 & sin(\alpha) \\ 0 & 1 & 0 \\ -sin(\alpha) & 0 & cos(\alpha) \end{bmatrix}$$

and

$$R_z(\alpha)=\begin{bmatrix} cos(\alpha) & sin(\alpha) & 0 \\ -sin(\alpha) & cos(\alpha) & 0 \\ 0 & 0 & 1 \end{bmatrix}$$

The FORTRAN coding of the four-fold rotation

$$H=R_z(\alpha)*R_y(\beta)*R_z(\gamma)*R_y(\delta)$$

resulting from symbolic computation might look like:

```
H11=-SIN(A)*SIN(C)*COS(D)-SIN(B)*SIN(D)*COS(A)+COS
(A)*COS(B)*COS(C)*COS(D)
H12=SIN(A)*COS(C)+SIN(C)*COS(A)*COS(B)
H13=-SIN(A)*SIN(C)*SIN(D)+SIN(B)*COS(A)*COS(D)+SIN
```

```
     . (D)*COS(A)*COS(B)*COS(C)
       H21=SIN(A)*SIN(B)*SIN(D)-SIN(A)*COS(B)*COS(C)*COS(
     . D)-SIN(C)*COS(A)*COS(D)
       H22=-SIN(A)*SIN(C)*COS(B)+COS(A)*COS(C)
       H23=-(SIN(A)*SIN(B)*COS(D)+SIN(A)*SIN(D)*COS(B)*
     . COS(C)+SIN(C)*SIN(D)*COS(A))
       H31=-(SIN(B)*COS(C)*COS(D)+SIN(D)*COS(B))
       H32=-SIN(B)*SIN(C)
       H33=-SIN(B)*SIN(D)*COS(C)+COS(B)*COS(D)
```

After passing the code through the Optimizer [15] for common subexpression extraction, the resulting code becomes:

```
G01  = SIN(A)
G02  = SIN(C)
G03  = COS(D)
G04  = SIN(B)
G05  = SIN(D)
G06  = COS(A)
G07  = COS(B)
G08  = COS(C)
G11  = G03*G07
G10  = G11*G08
G15  = G01*G02
G16  = G04*G05
G18  =  - G10 + G16
H11  =  - G18*G06 - G15*G03
G14  = G02*G06
H12  = G14*G07 + G01*G08
G12  = G05*G08
G09  = G12*G07
G13  = G03*G04
G17  = G09 + G13
H13  = G17*G06 - G15*G05
H21  =  - G14*G03 + G18*G01
H22  = G06*G08 - G15*G07
H23  =  - G14*G05 - G17*G01
H31  =  - G05*G07 - G13*G08
H32  =  - G02*G04
H33  = G11 - G12*G04
```

We process the code further by applying the following global operations:

-1- *Expression length reduction*: We assume here that the target machine does not allow chaining, such as the Cyber-205 [9]. Then shortening the expression length to a maximum of two operands can be attempted. For a Cray-1 this could be relaxed to longer expressions provided they produce no functional unit conflicts [16].

-2- *Topological sorting* to regroup statements within data dependency constraint boundaries. (This has been touched upon by various authors. See for instance [17]).

These two steps produce the following code with dependency boundaries explicitly shown.

```
      G01=SIN(A)
      G02=SIN(C)
      G03=COS(D)
      G04=SIN(B)
      G05=SIN(D)
      G06=COS(A)
      G07=COS(B)
      G08=COS(C)
C     -----------------
      G11=G03*G07
      G15=G01*G02
      G16=G04*G05
      G14=G02*G06
      G22=G08*G01
      G12=G08*G05
      G13=G03*G04
      G27=G08*G06
      G31=-G07*G05
      H32=-G02*G04
C     -----------------
      G10=G08*G11
      G20=-G03*G15
      G21=G07*G14
      G09=G07*G12
      G24=-G05*G15
      G25=-G03*G14
      G28=-G07*G15
      G29=-G05*G14
      G32=-G08*G13
      G33=-G04*G12
C     -----------------
      G18=-G10+G16
      H12=G21+G22
      G17=G09+G13
      H22=G27+G28
      H31=G31+G32
      H33=G11+G33
C     -----------------
      G19=-G06*G18
      G23=G06*G17
      G26=G01*G18
      G30=-G01*G17
C     -----------------
      H11=G19+G20
      H13=G23+G24
```

```
      H21=G25+G26
      H23=G29+G30
C     ------------------
```

-3- *Structure recognition* within dependency boundaries. In this step common structures of each type are detected and grouped together within dependency boundaries. This is done by projecting the original incidence matrices onto new matrices for common structures, as described in section 2 and in [11] and [18]. In the example above, negation of a simple product is recognized as a linked triad. At the end of this stage the representation of *the code is reverted to prefix form.*

-4- *Vectorization.* Variable names are renamed to display vector structure. The following auxiliary vectors are created:

-a- The *output vector*: This vector is associated with the sequence of statement names in the block (variables occuring on the left hand side of assignment statements) and is augmented by the list of auxiliary data elements such as input data and constants occuring in the block.

-b- The *input vectors*: The operands on the right hand side are associated with a number of help vectors having the same length as the output vector.

-c- The *gather vectors*: Renaming across dependency boundaries is recorded in these vectors. They appear in the gather-scatter function calls which are issued at code-output time and are known at load-time.

The resulting vectorized code in prefix form is shown below. The form is similar to the FOR Macro-reference implementation of REDUCE.

```
(NIL (VGATHR VOUT VG1 VIN1 5 12))
(FOR J (5 1 8) DO ((VOUT J) (SIN (VIN1 J))))
(FOR J (9 1 12) DO ((VOUT J) (COS (VIN1 J))))
(NIL (VGATHR VOUT VG1 VIN1 13 22))
(NIL (VGATHR VOUT VG2 VIN2 13 22))
(FOR J (13 1 20) DO ((VOUT J) (TIMES (VIN1 J) (VIN2 J))))
(FOR J (21 1 22) DO ((VOUT J) (MINUS (TIMES (VIN1 J) (VIN2 J)))))
(NIL (VGATHR VOUT VG1 VIN1 23 32))
(NIL (VGATHR VOUT VG2 VIN2 23 32))
(FOR J (23 1 25) DO ((VOUT J) (TIMES (VIN1 J) (VIN2 J))))
(FOR J (26 1 32) DO ((VOUT J) (MINUS (TIMES (VIN1 J) (VIN2 J)))))
(NIL (VGATHR VOUT VG1 VIN1 33 38))
(NIL (VGATHR VOUT VG2 VIN2 33 38))
```

```
(FOR J (33 1 33) DO ((VOUT J) (PLUS (MINUS (VIN1 J)) (VIN2 J))))
(FOR J (34 1 38) DO ((VOUT J) (PLUS (VIN1 J) (VIN2 J))))
(NIL (VGATHR VOUT VG1 VIN1 39 42))
(NIL (VGATHR VOUT VG2 VIN2 39 42))
(FOR J (39 1 40) DO ((VOUT J) (MINUS (TIMES (VIN1 J) (VIN2 J)))))
(FOR J (41 1 42) DO ((VOUT J) (TIMES (VIN1 J) (VIN2 J))))
(NIL (VGATHR VOUT VG1 VIN1 43 46))
(NIL (VGATHR VOUT VG2 VIN2 43 46))
(FOR J (43 1 46) DO ((VOUT J) (PLUS (VIN1 J) (VIN2 J))))
```

-5- *Output of code in target language.* The vectorized code can now be translated from prefix form to Fortran77 as is shown below. This could be done, for instance, by suitably interfacing to GENTRAN.

```
      REAL VOUT(46),VIN1(46),VIN2(46)
      INTEGER VG1(46),VG2(46)
      DATA VG1/1,2,3,4, 1,3,2,4,4,1,2,3, 9,5,7,6,12,12,
     .9,12,11,6,12,11,11,9,8,9,11,8,12,7, 23,24,25,20,
     .21,13, 10,5,10,5,39,41,28,30/
      DATA VG2/0,0,0,0, 0,0,0,0,0,0,0,0, 11,6,8,10,5,8,
     .7,10,8,7,13,16,18,14,14,16,14,16,19,18, 15,17,19,
     .29,31,32,33,35,35,33, 26,27,42,40/
C
      CALL VGATHR(VOUT,VG1,VIN1,5,12)
      DO 10 J=5,8
10      VOUT(J)  = SIN(VIN1(J))
      DO 20 J=9,12
20      VOUT(J)  = COS(VIN1(J))
C
      CALL VGATHR(VOUT,VG1,VIN1,13,22)
      CALL VGATHR(VOUT,VG2,VIN2,13,22)
      DO 30 J=13,20
30      VOUT(J) =VIN1(J)*VIN2(J)
      DO 40 J=21,22
40      VOUT(J) =-VIN1(J)*VIN2(J)
C
      CALL VGATHR(VOUT,VG1,VIN1,23,32)
      CALL VGATHR(VOUT,VG2,VIN2,23,32)
      DO 50 J=23,25
50      VOUT(J) =VIN1(J)*VIN2(J)
      DO 60 J=26,32
60      VOUT(J) =-VIN1(J)*VIN2(J)
C
      CALL VGATHR(VOUT,VG1,VIN1,33,38)
      CALL VGATHR(VOUT,VG2,VIN2,33,38)
      DO 70 J=33,33
70      VOUT(J) =-VIN1(J)+VIN2(J)
      DO 80 J=34,38
80      VOUT(J) =VIN1(J)+VIN2(J)
C
      CALL VGATHR(VOUT,VG1,VIN1,39,42)
      CALL VGATHR(VOUT,VG2,VIN2,39,42)
```

```
      DO 90 J=39,40
90      VOUT(J) =-VIN1(J)*VIN2(J)
      DO 100 J=41,42
100     VOUT(J) =VIN1(J)*VIN2(J)
C
      CALL VGATHR(VOUT,VG1,VIN1,43,46)
      CALL VGATHR(VOUT,VG2,VIN2,43,46)
      DO 110 J=43,46
110     VOUT(J) =VIN1(J)+VIN2(J)
```

Note that the necessary gather/scatter vectors VG1 and VG2 can be initialized at load-time. The arguments to the gather subroutine, VGATHR, are the source vector, the gather vector, the target vector, and the range indices, respectively. (The input data A,B,C & D are associated with the first four elements of VOUT; the output matrix H is imbedded in VOUT as well.) Note that in a multiprocessor architecture DO-loops occuring within a given dependency boundary can be executed in parallel. This is similarly true for the gather-calls. The code can now be expressed in Fortran 200 [9] with vector-syntax extensions:

```
REAL VOUT(46),VIN1(46),VIN2(46)
INTEGER VG1(46),VG2(46)
DATA VG1/1,2,3,4,  1,3,2,4,4,1,2,3,  9,5,7,6,12,12,
*9,12,11,6,12,11,11,9,8,9,11,8,12,7,  23,24,25,20,
*21,13, 10,5,10,5,39,41,28,30/
DATA VG2/0,0,0,0,  0,0,0,0,0,0,0,0,  11,6,8,10,5,8,
*7,10,8,7,13,16,18,14,14,16,14,16,19,18,  15,17,19,
*29,31,32,33,35,35,33, 26,27,42,40/
 VIN1(5;4)=Q8VGATHR(VOUT(1;46),VG1(5;8);VIN1(5;8))
 VOUT(5;4)=VSIN(VIN1(5;4);VOUT(5;4))
 VOUT(9;4)=VCOS(VIN1(9;4);VOUT(9;4))

 VIN1(13;10)=Q8VGATHR(VOUT(1;46),VG1(13;10);VIN1(13;10))
 VIN2(13;10)=Q8VGATHR(VOUT(1;46),VG2(13;10);VIN2(13;10))
 VOUT(13;8)=VIN1(13;8)*VIN2(13;8)
 VOUT(21;2)=-VIN1(21;2)*VIN2(21;2)

 VIN1(23;10)=Q8VGATHR(VOUT(1;46),VG1(23;10);VIN1(23;10))
 VIN2(23;10)=Q8VGATHR(VOUT(1;46),VG2(23;10);VIN2(23;10))
 VOUT(23;3)=VIN1(23;3)*VIN2(23;3)
 VOUT(26;7)=-VIN1(26;7)*VIN2(26;7)

 VIN1(33;6)=Q8VGATHR(VOUT(1;46),VG1(33;6);VIN1(33;6))
 VIN2(33;6)=Q8VGATHR(VOUT(1;46),VG2(33;6);VIN2(33;6))
 VOUT(33;1)=-VIN1(33;1)+VIN2(33;1)
 VOUT(34;5)=VIN1(34;5)+VIN2(34;5)

 VIN1(39;4)=Q8VGATHR(VOUT(1;46),VG1(39;4);VIN1(39;4))
 VIN2(39;4)=Q8VGATHR(VOUT(1;46),VG2(39;4);VIN2(39;4))
```

```
VOUT(39;2)=-VIN1(39;2)*VIN2(39;2)
VOUT(41;2)=VIN1(41;2)*VIN2(41;2)

VIN1(43;4)=Q8VGATHR(VOUT(1;46),VG1(43;4);VIN1(43;4))
VIN2(43;4)=Q8VGATHR(VOUT(1;46),VG2(43;4);VIN2(43;4))
VOUT(43;4)=VIN1(43;4)+VIN2(43;4)
```

4. Conclusions and Future Work

The example, given in the previous section, illustrates the vectorization of scalar block code by means of bottom-up structure recognition. We have addressed ourselves to a type of code typically resulting from algebraic processing. As this example is a small problem, the vectorized code will not necessarily speed up the computation (especially if one can recognize the matrix nature of the original problem). This is mainly due to the occurence of short vectors producing execution times which are characterized by start-up overhead. Furthermore the obligatory gathering of vectors to provide a constant stride increases the throughput time by a multiple of the vector length equal to the number of operands in a given vector expression. The problem of short vectors can however be resolved by inhibiting unprofitable vectorization. We are now experimenting further with a wider variety of strategies and larger problems. We point out again that the procedure illustrated here can also be applied to a large class of parallel architectures. Besides the parallel execution of independent DO-loops already mentioned, the gather/scatter operations could for instance map onto shared/local memory transfers.

Acknowledgement

We would like to thank the Dutch Working Group on the Use of Supercomputers for providing computer time with financial aid from the NWO and SURF foundations.

References

[1] Hearn, A.C. (1987). *REDUCE User's Manual Version 3.3*. The Rand Corporation.

[2] Gates, B.L. (1986). GENTRAN: An Automatic Code Generation Facility for REDUCE, *Proceedings SYMSAC '86* (Ed. B.W. Char), 94-99. ACM, New York.

[3] Broughan, K.A. (1986). A Symbolic Numeric Interface for the NAG Library, *NAG Newsletter* 2/86, 16-24.

[4] van den Heuvel, P., van Hulzen, J.A., Goldman, V.V. (To appear). Automatic Generation of FORTRAN-Coded Jacobians and Hessians, *Proceedings EUROCAL '87* (Ed. J.H. Davenport). Springer Verlag, Heidelberg.

[5] Hearn, A.C. (1985). Structure: The Key to Improved Algebraic Computation, *Proceedings RSYMSAC2* (Ed.s N. Inada & T. Soma), 215-230. World Scient. Publ., Singapore.

[6] van Hulzen, J.A. (1987). Program Generation through Symbolic Processing, *Essays on Concepts, Formalisms, and Tools* (Eds P.R.J. Asveld & A. Nijholt), CWI Tract 42, 231-260. CWI, Amsterdam.

[7] Arnold, C.N. (1983). Vector Optimization on the Cyber 205, *Proceedings of the Int. Conf. on Parallel Processing* (Eds H.J. Siegel & L. Siegel), 530-536. IEEE Comp. Soc. Press, Washington D.C.

[8] Kuck, D.J., Kuhn, R., Padua, D. Leasure, B. and Wolfe, M. (1981). Dependence Graphs and Compiler Optimizations, *Proceedings of the 8th ACM Symposium on Principles of Programming Languages*, 207-218. ACM, New York.

[9] CDC (1982). *Cyber Fortran 200, Version 1, Reference Manual*, Pub. 60480200, CDC, St. Paul, Minn.

[10] van Hulzen, J.A. (1983). Code Optimization of Multivariate Polynomial Schemes: A Pragmatic Approach, *Proceedings EUROCAL '83* (Ed. J.A. van Hulzen), Springer LNCS-series nr. 162, 286-300.Springer Verlag, Heidelberg.

[11] de Boer, J.K. (1986). *Bottom-Up Structure Recognition*, Master's Thesis (in Dutch), Twente University , Enschede, The Netherlands.

[12] Wang, P.S. (1986). FINGER: A Symbolic System for Automatic Generation of Numerical Programs in Finite Element Analysis, *J. Symb. Comp.*, 2, 305-316.

[13] Aho, A.V., Sethi, R., Ullman, J.D. (1986). *Compilers Principles, Techniques and Tools*. Addison-Wesley, Reading, Mass.

[14] Kuck, D.J. , Kuhn, R.H.,Leasure, B. and Wolfe, M. (1980). The Structure of an Advanced Vectorizer for Pipelined Processors, *Proceedings COMPSAC 80, The 4th International Computer Software and Applications Conf.*, 709-715.

[15] van Hulzen, J.A. and Hulshof, B.J.A. (In preparation). A Code Optimization Package for REDUCE.

[16] CRAY Research, Inc. (1986). *Fortran (CFT) Reference Manual*, Publication Nr. SR-0009, Mendota Heights, Minn.

[17] Bird, P.L. (1985). Linear Time Detection and Inherent Paralellism in Sequential Programs, *Computer Physics Comm.*, 37, 69-76.

[18] de Boer, J.K. , Goldman, V.V., van Hulzen, J.A. (In preparation). Bottom-Up Structure Recognition in Arithmetic Expressions.

Factorization in a Functional Language

A. C. Norman and J. Mitchell

Trinity College and Queens' College Cambridge

Introduction

One plausible hypothesis about parallel computations in application areas such as Computer Algebra where vector processing does not seem natural is that functional programming languages can provide the key that will unlock latent concurrency in programs and algorithms. This hope is based on both the fact that a functional programming language has semantics that do not depend on evaluation order (including the possibility of concurrent evaluation), and on the functional style seen in the 'single-assignment' languages developed on pragmatic grounds for use with the dataflow style of parallel computers. The work described here is an experiment to see how much trouble there might be re-implementing central parts of an algebra system using a pure side-effect-free programming style. The component of an algebra system selected for implementation in this style was the univariate part of a factorization package. This seemed both small and self contained enough to be managable, but sufficiently elaborate to reveal some of the difficulties of living without update-in-place operations and sequential computation. Programming was carried out in the language ML, using the Edinburgh implementation. This provided us with an opportunity to see how the strong (polymorphic) type system provided with ML felt in this application area as compared with the flexible type anarchy seen in a Lisp-coded factorizer one of us had worked on earlier. Although a working factorizer was produced, many of our experiences were negative: in

COMPUTER ALGEBRA AND PARALLELISM ISBN 0-12-221100-6

several areas where measurements on the Reduce factorization code had revealed performance hot-spots the restriction of a pure functional language seemed to impose inherent extra complications and cost, and the ML implementation used felt very slow in absolute terms even though it was run on a fast machine. The program structure imposed by ML, together with an attempt to produce a well-structured modular implementation of the factorizer package, left us with code which drew our attention very directly to the need for new debugging and program maintainance paradigms for use with functional code. In practise although the final code we produced does not rely on evaluation order effects or assignment operators, while testing it we found ourselves effectively forced into using some. Despite this we feel that quite modest achieved parallelism could recover the essential performance loss we see, better implementation technology would deal with the remainder of our concerns about speed and space, while program development tools designed to support functional and parallel computations, when they exist, will make testing and monitoring of code managable.

The Polynomial Manipulation Package

The starting point for our package was a body of code for manipulating polynomials. Despite the fact that the factorizer to be written later was only going to cope with the univariate case the polynomial package was made capable of dealing with multivariate expressions using a standard recursive datastructure strongly remeniscent of that used by Reduce. Polynomial addition, multiplication and division code up naturally with hardly any place where assignment statements or sequential processing could feel even remotely useful, and ML provides for **union** datatypes so having a polynomial as either an integer or a composite object consisting of a variable, power, leading coefficient and reductum was all very easy. One interpretation of what is meant by a pure functional language includes (in Lisp terms) the idea that 'eq' equality is not revealed, only 'equal' tests are permitted. Taking this line, of course, allows an implementation to copy datastructures (for instance between separate parts of the machine's memory) and make multiple copies as it sees fit. It also means that there it is not valid to think about saving space or time by trying to keep parts of datastructures stored uniquely (e.g. having all instances of the x raised to the fourth power referring to the same data object for (x^4)). As well as the (fairly trivial) code for polynomial arithmetic, our polynomial package contains code to read and print formulae. In the functional context we take 'reading' to mean accepting a linear list of characters and parsing them to yield a polynomial, while 'printing' has to perform the converse and map the nested structure of a polynomial onto a flat list of characters. Our reader view its input as a stream of characters and applies a series of filters to it. The first of these turns the stream of characters into a stream of tokens (each of which has to be badged with its type, e.g. number, operator, of expressions. There are respects in which the construction of this parser made us more aware than we had been of the interplay between parsing code and its effect on a notional cursor running over the input, and of the way in which bottom-up parsing techniques can be viewed as taking a succession of passes over input text producing versions of it where terminal symbols have been coallesced into relevant non-terminals.

Kroneker Factorization

Given a polynomial package our next experiment was to implement Kroneker's factorization algorithm. Although ML is a strict language we observed that for this algorithm there were several places where lazy evaluation would lead to a neater presentation of the complete factorization process. We therefore wrote the code with this evaluation discipline in mind, and later made a few changes so that the eager evaluation carried out by ML did not lead to disaster. Kroneker's algorithm involves selecting a collection of points at which the input polynomial is to be evaluated. All integer divisors of the values at these points are computed, and candidate factors are generated by interpolating through sets of divisors. The control structure for this search can feel ungainly when expressed in a sequential language. For our functional implementation we wrote code that described the generation of complete lists of divisors followed by the creation of a complete list of candidate divisor-sets. The final part of the code then just searches down this (very long!) list until it finds itself able to interpolate an integer-coefficient polynomial through the divisors, and until this polynomial is found to be a factor of the input. With lazy evaluation it is then automatically the case that divisor combinations are only generated when they are needed for testing, and indeed that the factorization of values of the input polynomial will not have to be completed if luck results in a factor being found early enough. Of course the main trouble with Kroneker's method is that when its input polynomial is irreducible it will have to consider **all** possible combinations of divisors, and in that case lazy evaluation is no great help.

Modular Arithmetic and Polynomials mod p

In order to implement the Berlekamp/Hensel factorizer we needed to provide support for arithmetic modulo various primes and prime powers. This is an area where most existing code will have made essential use of an updatable non-local variable: suppose it is called p, then using the ML syntax for an updatable cell one could go

```
val p = ref 3;
fun setmod n = (p := n);
fun modular_plus m n =  (m + n) mod !p;
```

and so on. (in ML the notation !p obtains the contents of the updatable cell refered to by p). This is not possible in the functional subset of ML that we wanted to restrict ourselves to, and the best scheme we could produce involved higher order functions:

```
fun make_mod_arith(p:int) =
  let fun mod_plus m n =  (m + n) mod p;
  ...
  in (mod_plus, mod_times, mod_div, ...) end;
```

The function make_mod_arith returns as its result a tuple containing functions which perform arithmetic operations modulo the specified value. Separate calls to make_mod_arith will return tuples of functions that are associated with different saved environments and hence different values of p. Now the polynomial manipulation package has to be rewritten as

```
fun make_poly_arith domain =
  let val (add, mult, ...) = domain;
  fun poly_plus ...
  ...
  in (poly_plus, ...) end;
```

where make_poly_arith expects to be handed a collection of functions as produced by make_mod_arith (or as created by hand to support non-modular arithmetic), and it yields an analagous set of polynomial-manipulating functions. There is a respect in which this structure seems elegant, and it could be seen as looking towards the (much more thoroughly worked out) mechanisms built into the new Scratchpad for creating new domains. But on the other hand it does seem to build in inflexible requirements on the set of arithmetic operators implemented in a domain, to be a way of slowing things down, and to demand that (for instance) the polynomial package be coded as an indivisible whole protected from testing and debugging by the wrapping of make_poly_arith. If ML-style compile-time typing is to be preserved it also seems hard to find a good way to reconcile a polynomial package with a field (e.g. the integers mod a prime p) with polynomials over a ring (e.g. the integers).

Berlekamp's Algorithm

Berlekamp's algorithm has two main parts. The first sets up a certain matrix and performs classical operations of linear algebra to find a basis for its null space. The second part has much in common with the systematic search performed in Kroneker's algorithm, in that it uses these null vectors to generate candidates for the factors of the input polynomial and it checks these candidates until the factorization is complete. Apart from the fact that it calls for functions to find the GCD of pairs of polynomials mod p, and to make such polynomials monic, the second part of the algorithm does not need further discussion here. The first part, being a calculation using matrices, might be expected to fit very neatly onto a vector-processing parallel computer. The traditional emphasis on update-in-place operations on matrices and the fact that ML does not (directly) support any array datatype make this step more visible as a potential performance bottleneck in the functional code. Our first implementation of Berlekamp's algorithm represented vectors as lists, and matrices as lists of vectors. Random access to a matrix element thus involves scanning down these lists, and updating a single element requires copying unpleasantly large amounts of structure. We can identify several different ways of doing much better than this, but are not yet clear which leads to the best results, and at least observe that in the case of linear algebra the restrictions imposed by side-effect-free programming make simple and fundamental algorithms much harder to program well. The possible improved null-space codes we have thought about are (1) use of binary trees rather than linear lists to represent matrices (to achieve log-time access and update), (2) Quad-tree representations for matrices and (3) clustering sets of access and update operations together so that the representation of the array and the processing performed on it work together instead of being in conflict. For the moment we have contented ourselves with the observation that (1) together with classical algorithms leads to a slow-down factor of only log(input-polynomial-degree) over imperative programming styles, and so can be considered acceptable.

The Hensel Construction

We found that the Hensel construction could be coded following very much the style we had previously used in imperative code. This is basically because the operations performed only involve a small amount of state, and this can be coped with by passing it down as arguments in a tail-recursive call. Thus even allowing for the early detection of factors this code almost wrote itself. The main oddities in the code are that ML's strong type system prevents one from making puns between polynomials with integer coefficients that just happen to be in the relevent range and polynomials that are known to have modular numbers as coefficients: we viewed the places where this meant that we had to insert explicit conversion functions as (at a minimum) good for the purposes of having well documented code. It was perhaps at the stage that the Hensel code was being exercised that we most felt need for better development support tools to go with our functional language: this was mainly because it relied upon and thus exercised all the preceeding code.

Conclusions

As well as limits to our package that we imposed for ourselves (such as the consideration of only univariate polynomials) we had constraints imposed on us by features of the particular ML implementation we used. For instance, although it clearly new about the idea of big integers, multiple precision arithmetic was not implemented. As a result of this we were able to run the Berlekamp code on the factorization problem from Sigsam problem 7 (the degree-40 polynomial that splits into four degree-10 factors), but not the Hensel code because that rapidly led to arithmetic overflow. Although the code we finished up with was code in a pure functional style we found it critically important at various stages in the development of the code to be able to insert print statements within our functions so as to trace their behaviour. This was needed both for debugging and for monitoring performance: if we followed up on the work with an attempt to produce a factorization package properly engineered for speed as well as robustness we would have done very much more of this. Using a functional style changed the areas where a factorizer seemed most ugly to code: in earlier imperative code for

Kroneker's method the generation of candidate factors and their elimination was done by code where the two concerns were entangled, while using streams in a functional world cleaned that up substantially. On the other hand, the matrix work in Berlekamp's algorithm still seems most naturally expressed using nested **for** loops in a language akin to C.

Parallel Polynomial Operations in the Large Buchberger Algorithm

Herbert Melenk Winfried Neun
Konrad-Zuse-Zentrum für Informationstechnik Berlin
Heilbronner Str. 10
D-1000 Berlin 31
Federal Republic of Germany

1. Distributive polynomials and Buchberger's algorithm

In the context of Gröbner base calculations, the polynomials are in distributive representation. In contrast to the recursive representation, a distributive polynomial (DP) is a linear sequence of monomials (MN), each of which consists of a coefficient ($C \in$ *coefficent domain*) and a product of variable powers (PVP):

$$\begin{aligned} DP &= [MN_1, \ldots, MN_k] \\ &= [C_1 x_1^{i_{11}} \cdots x_n^{i_{1n}}, \ldots, C_k x_1^{i_{k1}} \cdots x_n^{1_{kn}}] \\ &= \sum_{j=1,n} C_j x_1^{i_{j1}} \cdots x_n^{i_{jn}} \end{aligned}$$

The monomials are ordered in a descending sequence by a global term ordering, e.g. inverse lexicographical ordering:

$$MN_p > MN_q \longleftrightarrow \exists k \leq n : (\forall l < k : i_{pl} = i_{ql}) \wedge i_{pk} > i_{qk}$$

COMPUTER ALGEBRA AND PARALLELISM ISBN 0-12-221100-6

The standard constructors and selectors are:

select leading monomial	$Lmon([MN_1, \ldots, MN_k]) \rightarrow MN_1$
select reductum	$Red([MN_1, \ldots]) \rightarrow [MN_2, \ldots]$
add leading monomial	$AddL(MN_0, [MNP1, \ldots])$ $\rightarrow [MN_0, MN_1, \ldots]$

With these operators a complete polynomial arithmetic can be constructed.

The Buchberger algorithm (Bba) constructs critical pairs of polynomials and processes them. In this context, we examine its central part in a (for description purposes) simplified version, which omits all details (e.g. vanishing polynomials, criteria, full reduction). The description of the full Bba can be found in Gebauer, Möller [88] and Möller [88].

```
while notempty pairlist do                              % main loop
    ≪ (p1, p2) := next pair;
    q := lcm(Lmon(p1), Lmon(p2));
    s := (q/Lmon(p1)) * p1 − (q/Lmon(p2)) * p2;
    h := s;
                                                         % reduction loop
    while exists r ∈ G such that Lmon(r) divides Lmon(h) do
           h := h − (Lmon(h)/Lmon(r)) * r;
    h := 1/(leadingC) * h;                               % normalizing
    for each r ∈ G construct pair (r, h)
           testing criteria on Lmon(r), Lmon(h);
    G := {h} ∪ G;≫
```

When we classify the polynomial operations needed by the Bba we find that:

(1.1) for the control of the algorithm only the access to the leading monomial is necessary,

(1.2) all arithmetic combinations are special cases of the pattern

$$MN_1 * DP_1 + MN_2 * DP_2$$

with appropriately constructed monomials.

These facts will be important for our approach to parallelization.

2. LCOMB: linear combination as a central tool for distributive polynomial operation

Because of (1.2) we introduce a new operation:

$$Lcomb(MN_1, DP_1, MN_2, DP_2) \rightarrow MN_1 * DP_1 + MN_2 * DP_2$$

which is a "linear combination" of two polynomials.

The Bba can be rewritten using Lcomb as single polynomial operation, but Lcomb can support the complete polynomial arithmetic as well:

$$\begin{aligned} &DP_1 + DP_2 = Lcomb(1, DP_1, 1, DP_2)\\ &DP_1 - DP_2 = Lcomb(1, DP_1, -1, DP_2)\\ &DP_1 * DP_2 = \ll x := 0;\\ &\quad \textit{for each } MN \textit{ in } DP_1 \textit{ do } x := Lcomb(1, x, MN, DP_2);\ x \gg \end{aligned}$$

The straightforward code for Lcomb is recursive, the depth determined by the length of the polynomials (omitting details regarding efficiency etc.).

$$\begin{aligned} &\textit{proc } Lcomb1(MN_1, DP_1, MN_2, DP_2): \qquad (2.1)\\ &\quad (C_1, PVP_1) := MN_1 * Lmon(DP_1); (C_2, PVP_2) := MN_2 * Lmon(DP_2);\\ &\quad \textit{if } PVP_1 = 0 \wedge PVP_2 = 0 \textit{ then } 0\\ &\quad \textit{else if } PVP_1 = PVP_2 \textit{ then}\\ &\quad\quad AddL\ ((C_1 + C_2, PVP_1), Lcomb(MN_1, Red(DP_1), MN_2, Red(DP_2)))\\ &\quad \textit{else if } PVP_1 > PVP_2 \textit{ then}\\ &\quad\quad AddL((C_2, PVP_2), Lcomb(MN_1, Red(DP_1), MN_2, DP_2))\\ &\quad \textit{else } AddL((C_2, PVP_2), Lcomb(MN_1), (DP_1, MN_2, RedDP_2))) \end{aligned}$$

Because of the recursion and of the constructor *AddL*, the leading monomial is added last. This technique is efficient for single processor LISP style systems. If we add a new constructor

$$\text{Append monomial: } App\ ([MN_1, \ldots, MN_k], MN_p) \rightarrow [MN_1, \ldots, MN_k, MN_p]$$

we can reformulate *Lcomb* in the style of a merge procedure

$$\begin{aligned} &ProcLcomb(MN_1, DP_1, MN_2, DP_2, r): \qquad (2.2)\\ &\textit{begin}\\ &\quad (C_1, PVP_1) := MN_1 * Lmon(DP_1); (C_2, PVP_2)\\ &\quad\quad := MN_2 * Lmon(DP_2); \end{aligned}$$

```
loop :
    if 0 = PVP1 = PVP2 then return r;
    if PVP1 = PVP2 then ≪ r := App(r,(C1 + C2, PVP1));
      DP1 := Red(DP1);(C1, PVP1) := MN1 * Lmon(DP1);
      DP2 := Red(DP2);(C2, PVP2) := MN2 * Lmon(DP2) ≫
    else if PVP1 > PVP2 then
    ≪ r := App(r,(C1, PVP1));
      DP1 := Red(DP1);(C1, PVP1 := MN1 * Lmon(DP1) ≫
    else
    ≪ r := App(r,(C2, PVP2));
      DP2 := Red(DP2);(C2, PVP2 := MN2 * Lmon(DP2) ≫;
goto loop;
end;
```

In this version the fifth parameter "r" has to be initialized by the caller as an empty polynomial; we will need this convention later.

3. Parallel execution based on Lcomb

In (2.2) the leading monomial is calculated first and it is the first monomial to be linked to the resulting structure. Because of (1.2) we can continue the Bba as soon as the leading monomial is available, so we can overlap the operation of *Lcomb* and the processing of its result by the Bba (including further Lcomb's) in a multi-processing environment provided that we have a synchronization regarding production/access of subsequent monomials.

Figure 3.1 demonstrates, that even several cycles of the main loop in Bba can operate simultaneously. Note that although the global structure of the algorithm looks very homogeneous, it cannot be configured as systolic array because the speed of the data flow through Lcomb is not constant: if the exponent patterns of the meeting monomials are equal, Lcomb consumes one monomial from both inputs, otherwise from one input alternating irregularly and there are steps where no output monomial is produced (coefficient extinction). The number of R-boxes between two S-boxes is at all not predictable.

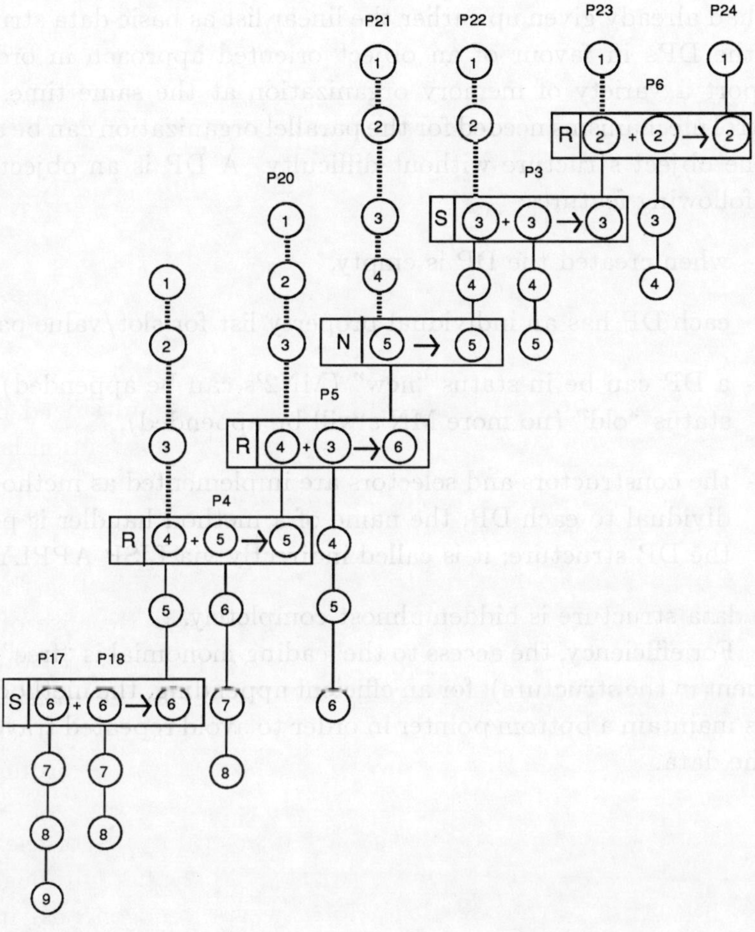

Figure 3.1: Snapshot of overlapped Lcomb actions:

S	:	calculation of an S-polynomial
R	:	reduction step
N	:	normalizing
P19	=	S(P17, P18)
P21	=	H(P17, P18)
P23	=	S(P22, P3)

4. Supporting Data Structure: Objects

We had already given up earlier the linear list as basic data structure for the DPs in favour of an object oriented approach in order to support a variety of memory organization at the same time. The control mechanisms needed for the parallel organization can be added to the object structure without difficulty. A DP is an object with the following features:

- when created the DP is empty,
- each DP has an individual property list for slot/value pairs,
- a DP can be in status "new" (MN2's can be appended) or in status "old" (no more MN's will be appended),
- the constructors and selectors are implemented as methods individual to each DP; the name of a method handler is part of the DP structure; it is called indirectly via LISP APPLY.

The data structure is hidden almost completely.

For efficiency, the access to the leading monomial is "free" (first element in the structure); for an efficient appending, the method handlers maintain a bottom pointer in order to avoid repeated traversing of the data.

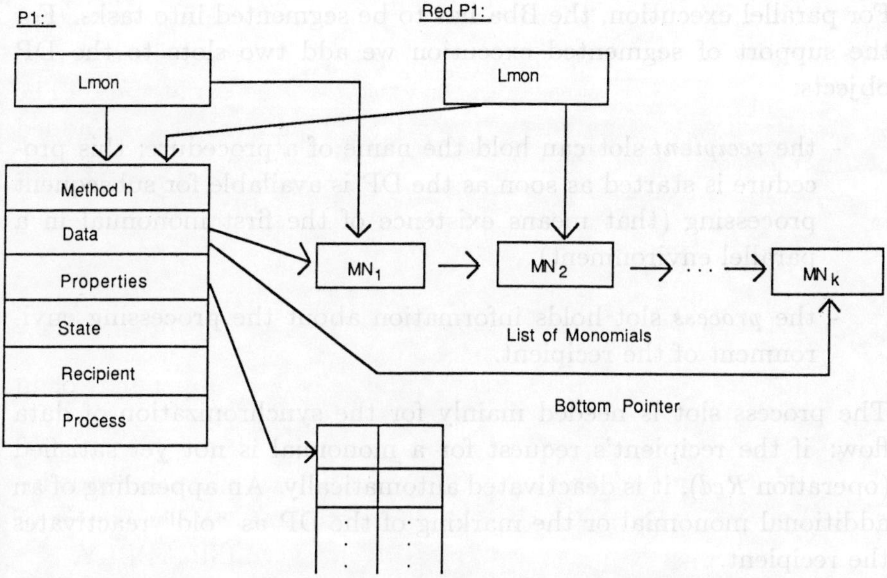

Figure 4.1: Object structure of a DP "P1" and of its reductum.

The head of an object is implemented as list structure. The operation *Red* produces a similar object with another first element and identical rest; the original object remains unchanged.

5. Segmentation of the Buchberger Algorithm

For parallel execution, the Bba has to be segmented into tasks. For the support of segmented execution we add two slots to the DP objects:

- the *recipient* slot can hold the name of a procedure; this procedure is started as soon as the DP is available for subsequent processing (that means existence of the first monomial in a parallel environment).

- the *process* slot holds information about the processing environment of the recipient.

The process slot is needed mainly for the synchronization of data flow: if the recipient's request for a monomial is not yet satisfied (operation *Red*), it is deactivated automatically. An appending of an additional monomial or the marking of the DP as "old" reactivates the recipient.

With this technical background we can rewrite the Bba in a segmented form (sBba), which enables overlapped processing. The routine $Pcreate\ (fnc(P_1, \ldots, P_n))$ creates a new process, which performs the function "fnc" with parameters $P_1, \ldots, P_n$. In most cases the last parameter is an object with defined recipient; as soon as the first monomial arrives, the object creates a process with this recipient as function and itself as parameter.

```
proc Bba1():          % calculate S-polynomial
    if empty pairlist then Pcreate (Finale) else
    ≪ (P1P2) := next pair;
       q := lcm(Lmon(P1), Lmon(P2));
       Pcreate (Lcomb(q/Lmon(P1), P1, (-q)/Lmon(P2), P2,
                new DP (recipient = Bba2))
    ≫;

proc Bba2(s):         % reduce S-polynomial to H-polynomial
    if exists r ∈ G such that Lmon(r) divides Lmon(s)
      then            % reduction step (loop)
          Pcreate (Lcomb(-1, s, Lmon(s)/Lmon(r), r,
```

```
                new DP(recipient = Bba2)
     else       % normalization step (exit from loop)
           Pcreate (Lcomb(0, 0, 1/(leading C of S), S,
                new DP(recipient = Bba3);

proc Bba3(h):   % construct new pairs, add h to G, loop
   ≪ for each r ∈ G  construct pair (r, h)
             testing criteria on (Lmon(r), Lmon(h);
     G := {h} ∪ G;
     Pcreate (Bba1());
   ≫ ;
```

This version of the Bba is performed as a chain of coroutines:

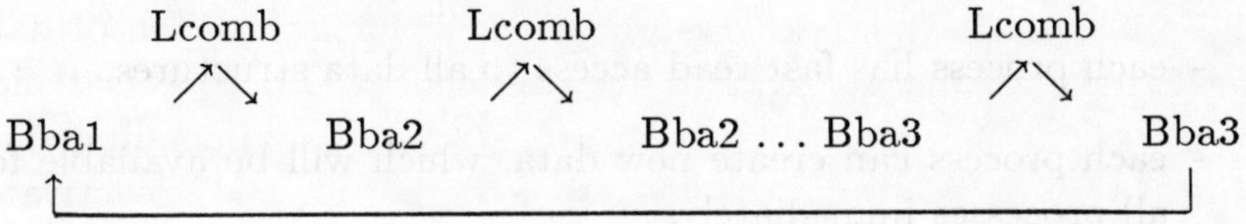

The routines Bba1 - Bba3 control the algorithm and predetermine in the recipient slot of the resulting structure the successor of an Lcomb action.

6. Handling of mixed data flow via closures

In (5) all processes had the simple structure one/two polynomials in - one polynomial out. If a data structure has to traverse several processes, this can be formulated by means of Common LISP type closures. As an example, the parallel version of the general multiplication of two DPs can be formulated with a closure as recipient, which simulates the loop over DP1:

```
proc multx(DP1, DP2, DPr):
     if DP1 = 0  then Pcreate (Result(DPr))
        else
        Pcreate (Lcomb(Lmon(DP1), DP2, 1, DPr,
              new DP(recipient = #'(λ(r)(DP1 := Red(DP1);
              multx(DP1, DP2, r))));
```

It is activated with an empty DP (with or without recipient) as third parameter.

In a similar manner the reduction loop of Bba can be coded, if full reduction (that is reduction of monomials behind the first "stable" Lmon of an H-polynomial) is required.

7. Process Model

The parallel execution requires a technical base, which can handle processes with the following features:

- each process has the full LISP(REDUCE) functionality without I/O functions,
- each process has fast read access to all data structures,
- each process can create new data, which will be available for all processes immediately,
- there is a synchronizing mechanism for the safe accessing/updating of critical data,
- a process can wait for an event (creation of data in a specific structure).

There are several different implementations of the last feature possible; the simplest will be a "busy waiting" loop, the best would be a multi-processing environment: the processor sets a waiting process asleep and executes another active process instead.

An always important question in parallel processing is the granularity of the decomposition. In this approach the "data grain" is the DP, assuming for a moment that the data flow is steady. In a complicated calculation the "typical" intermediate DP has dozens of monomials with quotients of big integers as coefficients.

The overall processing is asynchronous in nature. A forced synchronization only takes place in case of monomial extinction. So the case of a zero H-polynomial is a local fence for the parallel execution.

8. Simulation

For the first implementation of sBba a multi-processing systems was simulated in a conventional LISP environment. For simplification we assume that all loop steps of Lcomb processing one monomial need the same computing time $t = 1$ and that Bba1, Bba2 and Bba3 too need the time $t = 1$ for execution. The number of processors available is a parameter which can be set to only value between 1 and ∞.

There are three queues:

Q_e executing processes
Q_w processes waiting for execution
Q_s sleeping processes

At the beginning of each time step t Q_e is filled up from Q_w up to the number of processors available. Then each element of Q_e is allowed to execute one step. If a process terminates, it is deleted from the Q's. If a process needs data not yet available, it moves from Q_e to Q_s. If data arrive, it moves from Q_s to Q_w. Each process executing in time step t with a record (process#, t#) to a log file. This log file allows the evaluation of parallelism in the execution. As an example we give a graphical representation of the execution of the well-known "little Trinks" (Böge et al. [86]) with different numbers of processors. This example is a non-typical small one, but it demonstrates the depending length of polynomials (= number of time steps) and parallelism.

```
f1 := 45*p + 35*s -165*b -36;
f2 := 35*p + 40*z + 25*t - 27*s;
f3 := 15*w + 25*p*s +30*z -18*t -165*b**2;
f4 := -9*w + 15*p*t  + 20*z*s;
f5 := w*p + 2*z*t - 11*b**3;
f6 := 99*w - 11*s*b +3*b**2;
f7 := B**2 + 33/50*b + 2673/10000;
```

The input polynomials for the sample calculation.

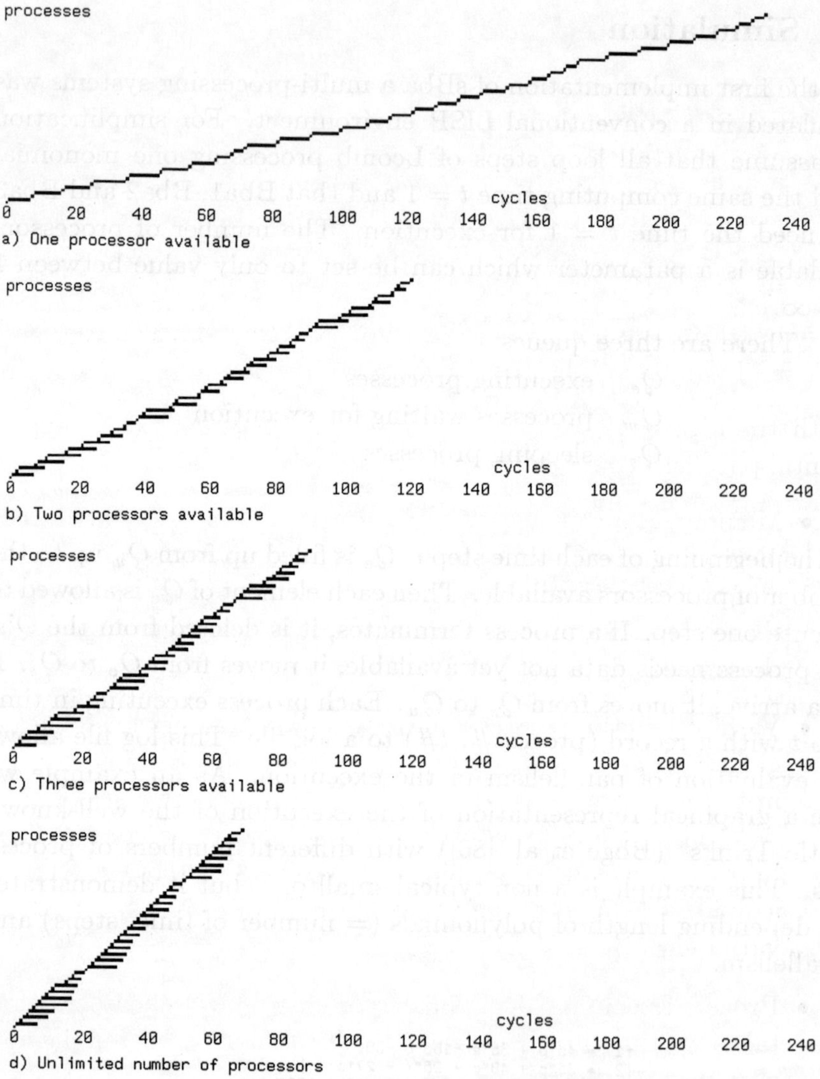

Figure 8.1: One calculation simulated with different numbers of processors. From (a) to (b) there is a "speed-up" of almost two. (d) shows that there is a close connection between the length of polynomials and the parallelism. The zero reductions are to be seen as bottlenecks.

9. Parallel execution of the Bba

The CRAY X-MP multiprocessor architecture supports parallel execution by:

- a large central memory directly accessible by all processors,

- sets of shared registers and semaphores (each application has its own set) with fast access,

- operating systems which allow an arbitrary mix of sequential and parallel applications in a multiprogramming environment.

With these prerequisites we designed and implemented an experimental parallel PSL with the following features:

- Memory Management: Code and data area are shared completely. There is only one heap; all processors are allowed to allocate data in the heap; the critical heap pointers reside in shared registers and are protected by a semaphore such that their updating is sequentialized.

- Variable Binding: PSL supports shallow binding for fluid and global variables. For efficiency we have adopted this technique in the following sense: each processor has its own set of value cells; fluid values are stored in these local value cells and will be rebounded when processes are exchanged (so the scope of a fluid variable is a calling tree rooted in its binding); global variables have a system wide unique value cell.

- Process Execution: Each processor has its own scheduler, which takes waiting processes from the queues and executes them. If a process is interrupted for missing data, it is deactivated and the scheduler selects another process for execution. The interrupt has an S-expression as parameter, which describes the reason for the interrupt in form of a test. So the evaluation of this test determines, if a process can be resumed.

- High Level Interface: Usually parallel execution in LISP is coded via the "future" feature [Swanson et al]. Futures are adequate to model the dynamic behaviour of DP's during construction, but those parts of the Bba working with sets of polynomials and critical pairs are not free of side effects, so they need additional primitives for synchronization. We decided to start with more elementary tools which later can be combined in higher levels:

pcreate (fn, pars) creates a new process as an asynchronous *apply*.

pinterrupt (test) interrupts an execution until the value of test will be T. In the Bba this primitive is called only in the most inner DP access routine.

plock (sem) requests exclusive access to a resource described by a user semaphore (global variable).

punlock (sem) releases exclusive usage.

Note that a request for an already locked semaphore causes an interrupt. The user semaphores are not identical with the hardware semaphores; the latter cause a processor to wait and they can be used for very short code sequences, e.g. for the implementation of a "test-and-set" of user semaphores.

There is no theoretical limit for the number of processors handled by this organization, although the concurrent heap management can become a bottleneck for larger numbers of processors. In the ideal case of no contention each processor operates with the full speed of the standard CRAY PSL.

First experimental calculation with the parallel Bba on a 2-processor X-MP were performed with the "Big Trinks" examples from [Böge et al.]. At the moment only the total cpu time (spent on both processors) was measured in order to get an impression of the overhead produced by the process management and the interrupts. All these tests were done in the standard workload which of course imposes additional interrupts by scheduling both processors freely among all user jobs.

conventional program with	3.6 sec
parallel program with 1 processor assigned	3.7 sec
parallel program with 2 processors assigned	3.7 sec - 4.5 sec (average 4.0 sec)

These tests will be continued with larger examples and a dedicated environment when the basic implementation reaches stable state.

10. Further Parallelism in the Bba

Until now the Buchberger algorithm itself remained untouched. All actions take place in the same order as described in [Gebauer, Möller]; the parallelism is based only on overlapping of steps in a context comparable to pipelining. If we denote with LCP the list of critical pairs and with t the number of h-polynomial calculation, the sequence of operations is:

select one critical pair from LCP(t)
calculate h(t+1) from this pair
update LCP(t) with h(t+1) giving LCP(t+1)

If LCP(t) has more than one element (which is true in most steps), more than one h-calculation can be started at the same time or, e.g. if a processor idles, a new pair can be selected from LCP(t) before LCP(t+1). This technique requires additional bookkeeping, but a higher degree of parallelism can be expected. The price for that parallelism is that a calculation could be started, which would be recognized obsolete during the next update of LCP. These effects will be object of further investigation.

References

Anderson, J. W., Galway, W. F., Kessler, R. R., Melenk, H., Neun, W. (1987). Implementing and Optimizing Lisp for the CRAY, IEEE Software.

Böge, W., Gebauer, R., Kredel, H. (1986). Some Examples for Solving Systems of Algebraic Equations by Calculating Groebner Bases, J. Symb. Comp. 1986, vol.2, pp. 86 – 98.

Buchberger, B. (1965). Ein Algorithmus zum Auffinden der Basiselemente des Restklassenrings nach einem nulldimensionalen Polynomideal, Ph.D. thesis, Innsbruck.

Buchberger, B. (1985). Gröbner bases: An algorithmic method in polynomial ideal theory. (In: Multidimensional Systems Theory, ed.: N. K. Bose), pp. 184 – 232., D. Reidel Publ. Comp.

Gebauer, R., Möller, H. M. (1988). On an installation of Buchberger's algorithm. To appear in J. Symb. Comp 1988.

Hearn, A. C. (1987). REDUCE User's Manual, Version 3.3. The Rand Corporation.

Möller, H. M. (1988). On the construction of Gröbner bases using syzygies. To appear in J.Symb. Comp. 1988.

Melenk, H., Neun, W. (1987). REDUCE User's Guide for CRAY 1/ X-MP Series Running COS, Technical Report TR 87-4. Konrad-Zuse-Zentrum für Informationstechnik Berlin.

Melenk, H., Neun, W. (1986). Usage of Vector Hardware for LISP Processing. Konrad-Zuse-Zentrum für Informationstechnik Berlin.

Swanson, M. R., Kessler, R. R., Lindstrom, G. (1988). An implementation of Portable Standard Lisp on the BBN Butterfly, TR 88-01. Department of Computer Science, University of Utah.

Implementation of a parallel algorithm to compute a Gröbner basis on boolean polynomials.

Pascale SENECHAUD[1]

Laboratoire TIM3-IMAG
Institut National Polytechnique de Grenoble
38031 GRENOBLE Cedex

Abstract: Gröbner bases are an important mathematic tool which has applications in geometry and algebra [Buchberger]. Some sequential algorithms to compute a Gröbner basis have been implemented in most of computer algebra systems (Macsyma,ScratchpadII) and also on Macintosh in C [Bayer and Stillman]. A parallelization of these algorithms, interesting because of the data size, has only been studied with different machine model from this presented here [Watt].

We study boolean polynomials for which the construction of Gröbner bases is more simple. The generalization does not modify the parallel algorithm structure: Only the basic operations and the data structure are different.

Boolean Gröbner bases have applications in logic [Chazarain].

I. / General Presentation

The definition and the way to compute a Gröbner basis is supposed to be known. None of the following notions will be developped: Critical pair, Spolynomial, normalisation. We want to parallelize the Buchberger 's

[1] This work is supported by *PRC Mathematiques et Informatique* and by the *Greco Calcul Formel* of the French *Centre National de la Recherche Scientique* (CNRS).

COMPUTER ALGEBRA AND PARALLELISM ISBN 0-12-221100-6

algorithm [Buchberger1].
A parallel algorithm is presented.We have determinated the independant tasks of the sequential algorithm and the proposed algorithm adds recursively adequate polynomials to the set of input polynomials. The fact that we work with boolean polynomials involves particular choices in the order of the variables, and simplifies the basic operations.

We work in $F_2 [x_1, \ldots, x_n] / (x_1^2 + x_1, \ldots, x_n^2 + x_n)$ where F_2 is the field $\mathbb{Z}/2\mathbb{Z}$ and where $(x_1^2+ x_1, \ldots x_n^2+ x_n)$ is the ideal generated by the polynomials $x_1^2+ x_1, \ldots x_n^2+ x_n$.
So we have the following properties:

i) $x_i^2 = x_i$ for all i in {1,..., n}.

ii) $x_i + x_i = 0$ for all i in {1,..., n}.

II / The parallel algorithm

We assume to have at one's disposal a parallel machine with n processors, each having a local memory. The processors can be connected in order to form a ring. The algorithm presented below has been implemented on the hupercube FPS T20 of the TIM_3 laboratory.

II.1 / computation of Spolynomials

The computation of the Spolynomials may be done in parallel since they are independant. If we have m polynomials in input and n (n is even) processors at one's disposal, with m ≥ n.We distribute the m polynomials among the processors' memories in the following way:

$$m = qn + r \quad (0 \leq r < q)$$

(n - r) processors contain q polynomials
the r remaining processors contain q+1 polynomials.

The main problem when computing the Spolynomialst hat is all the polynomials must meet each other.

i) In a first step, we compute in parallel, the Spolynomials associated to the polynomials contained in the memory of each processor. Let us suppose that k is the number of these polynomials
In a processor:
Beginning with $FP = \{ p_1, p_2, ...p_k \}$ we compute C_{12} ,...., C_{1k}, C_{23}, ..., C_{2k}, ...,C_{k-1k} where C_{ij} is the Spolynomial associated to p_i and p_j

ii)The subsets of input polynomials contained in each processor circulate along the ring which allows us to compute all the Spolynomial associated to the m input polynomials.

Let FP_i be the collection of k polynomials contained in the ith processor of the ring. During the first step we have computed, in each processor, the Spolynomials associated to these polynomials. We shall note FC_i the collection of Spolynomials obtained in the ith processor of the ring. We proceed as follow:

i) The collection FP_i is transferred from the processor i to the processor i+1.

ii) The Spolynomials $FC_{i,i+1}$ from FP_i and FP_{i+1} are computed in each processor.

iii) FC_i and $FC_{i,i+1}$ are concatenated in parallel .

These three steps are then repeated until we have computed m (m-1)/2 Spolynomials associated to the m input polynomials. With four processors this manipulation may be represented in the following way:

First iteration:

i^{th} processor of the ring ($i \neq 1$):

- $FP_i \rightarrow FC_i$

- send FP_i

- receive FP_{i-1}

- $(FP_i, PP_{i-1}) \rightarrow FC_{i,i-1}$

On a ring:

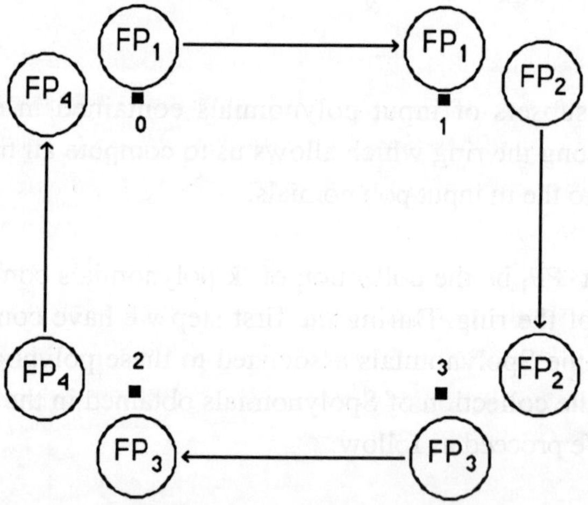

And so on, until having computed all the Spolynomials associated to the m input polynomials.

II.2 / The normalization

All the Spolynomials obtained must be normalized with respect to the input polynomials. The process is the same as during the computation of

the Spolynomials. Each processor contains two collections of polynomials: the collection of its k input polynomials, and a collection of Spolynomials, resulting from the former steps.
The Spolynomials circulate along the ring and are normalized with respect to the given polynomials as soon as they meet them in the processors.
Then, the processors work in parallel to reduce the Spolynomials they receive with respect to the collection of the input polynomials they contain respectively.
The execution of this step is finished when all the Spolynomials are in normal form. The processors stop the computation simultaneously.

II.3 / The whole algorithm

When the Spolynomials normalized are computed, in order to obtain a Gröbner basis, the sequential algorithm repeats the same computation, substituting, at each iteration, the collection of input polynomials by the union of this collection with the collection of the Spolynomials normalized computed at the former step.Therefore, we may repeat, in the parallel algorithm, the step of computation and normalization of Spolynomials substituting, at each step and in each processor i of the ring, FP_i by the concatenation of FP_i and $FPCN_i$. We use this method, but modified, because it generates, at each step results yet obtained in the former step.

III / Implementation on the FPS T20

III.1 / Choice of a representation for the boolean polynomials

The choice of the structure to represent the boolean polynomials is justified by the fact that we want to translate the basic operations, such as the sum and the product of boolean polynomials, into a simple manpulation of that structure. This choice conditions the order on the variables.

In a first time we have represented a boolean polynomial by an array of monomials, and each monomial by an integer. This integer is build (the integers are written in radix 2, with 32 bits) in the following way:

the monomial 1 is represented by 0................01 which is equal to 1

the monomial x_i is represented by 0......010....01 which is equal to $2^i + 1$

↑
i+1 th bit

the monomial $x_i x_j$ is represented by 0.....010...010...01 which is equal to $2^i + 2^j + 1$

With this representation, and since the boolean operations are available on the integers, we easily translate the necessary operations for the computation of a Gröbner basis.

In order to represent polynomials with more than 32 variables, we now work using large numbers [Roch].

III.2 / Results

The algorithm reads in input the number of polynomials, the number of processors wanted and the polynomials.

The following results show the evolution of the computing time of a Gröbner basis according to the number of processors.

The treatment of 32 polynomials with 5 variables gave us the following results:

num. of proc	time in sec
1	1.673
2	0.501
4	0.290
8	0.226
16	0.217

The time decreases as the number of polynomials increases. This decreasing is reduced by the delays of communications, which increase according to the number of processors . The more the computations are important in front of the communications the more the algorithm is interesting. It is difficult to control the number of simplifications required by the algorithm and then to know how the time of communication grows according to the data. So we are studying an other algorithm where the communication time does not depend on the simplifications. Instead of considering the independancy of certain tasks in the sequential algorithm, we use the following fact:

Let $P= (p_1,...., p_k)$ a set of polynomials.
Let $P_1 = (p_1,...p_p)$, $P_2 = P\text{-}P_1$ and G_1 and G_2 gröbner bases associated respectively to P_1 and P_2.
A Gröbner basis of the union the G_1 and G_2 is also a Gröbner basis associated to P.

We do not describe the corresponding algorithm here.

REFERENCES

[Bayer] D. Bayer and M. Stillman "The Design of Macaulay: A System for Computing in Algebraïc. Geometry and Commutative Algebra." (January 1986).

[Buchberger1] B. Buchberger. "A Criticical Pair / Completion Algorithm for Finited Generated Ideals in Rings". Proc logic and Machines. Decision Problems and Complexity ed by E. Bröger, G.Hasenjaeger, D. Rödding.Springer LNCS 171 (1983).

[Buchberger2] B. Buchberger. "Basic Features and Developpment of the Critical Pair / Completion Procedure".Preprint J.Kepler University Austria

[Chazarain] J. Chazarain. "The Lady, the tiger and the Gröbner Basis". Preprint n°100 University of Nice.Department of Mathematics.

[FPS] Floating Point Systems, "Programming the FPS T–Series, Release B, Portland Oregon 97223.

[Roch1 and al.] J.L.Roch, P.Sénéchaud, F.Siebert et G.Villard, "Parallel Algebraic Computing", Imag Grenoble, RR–686 I , (december 1987).

[Roch2 and al.] J.L. Roch, P. Senechaud, F. Siebert, G. Villard "Calcul Formel, Parallelisme et Occam" OPPT Ed.T.Muntean (1987).

[Watt] S. Watt. "Bounded Parallelism in Computer Algebra". Thesis presented to the University of Waterloo Ontario (1985).

ADAPTATION OF SAC-1 ALGORITHMS FOR AN SIMD MACHINE

Dennis Weeks
Convex Computer Corp.
1731 Technology Drive
San Jose, CA 95110

Abstract

The list processing and multiple-precision integer arithmetic procedures of the SAC-1 software package have been adapted for use on the Convex C-series, a family of "mini-supercomputers" with Cray-like vector registers. A 45:1 speedup was obtained on a specific test case of Buchberger's Groebner basis algorithm by "tuning" the SAC-1 routines for the Convex vector architecture. Although the vector architecture was the motivation for the changes, a close analysis reveals that a significant part of the improved performance was due to generic changes in the list-processing algorithms which would be applicable to almost any computer, and another major improvement was due to the Convex computer's ability to do 64-bit integer arithmetic.

SAC-1 is a collection of ANSI 1966 Fortran subroutines commonly used in symbolic algebraic computations. It is a mature product, nearly 20 years old, consisting of over a dozen functional groups of subroutines. The list processing and multiple-precision integer arithmetic routines [Collins 1971a,b] are incorporated into ACM Algorithm 628, a definitive published version of Buchberger's Groebner basis algorithm [Winkler *et al*, 1985]. The decision to use the SAC-1 codes was based on their high degree of portability and their long history of robustness.

Since SAC-1 was first introduced, however, there have been many advances in computer hardware and software. Some of these, such as faster execution speeds and compiler optimization techniques, require at most re-compilation in order to benefit existing programs. But in other cases the advances represent major architectural changes — such as instruction-level parallelism — whose applicability to existing programs is not always obvious. Compilers may try to

COMPUTER ALGEBRA AND
PARALLELISM ISBN 0-12-221100-6

restructure programs to use these features, but their approach is nearly always "bottom-up" and only a programmer who can take a "top-down" view of the entire program will be able to restructure the code in a way that will permit the compiler to produce "vector" or "parallel" code to implement the user's algorithm.

Because of the pervasiveness of the Buchberger algorithm in current research in computer algebra, it is desirable to have implementations of it that take full advantage of the underlying computer architecture. Its execution time increases rapidly (probably exponentially) in the degree and the number of variables of the input polynomials. In the absence of new algorithmic breakthroughs, faster execution speed is the only means by which problems that are "large enough to be interesting" can be solved in reasonable amounts of time.

I therefore decided to try to implement the Buchberger algorithm for the Convex C-series processors, to see how effective these computers would be for symbolic algebra codes. While supercomputers, like the Cray machines, are generally regarded as best suited for intensive floating-point computation, the Convex architecture also includes a full complement of vector instructions for integer arithmetic, so this seemed like a worthwhile exercise. The results, as described below, certainly justified the effort. (Melenk and Neun, elsewhere in these proceedings, report similar encouraging results on a Cray-1 despite the lack of integer division instructions, and having to implement vector multiplication of integers using floating-point multiplications.)

The study was conducted on a C-120, the smallest and slowest of the C-series processors. It is an SIMD (single instruction, multiple data) machine with eight vector registers, each having 128 64-bit elements. The primary means of improving performance is to do arithmetic in the vector registers, with the best results obtained when the vector length is 128. Further improvement is possible by overlapping vector addition/subtraction and multiplication/division operations. Since the SAC-1 procedures operate on lists rather than arrays of numbers, as published they contain no code that is immediately "vectorizable", and the main thrust of this investigation was to identify modifications that would permit frequent use of the vector registers. Although the vector registers were the motivation for this exercise, two other categories of changes were also found to have a significant impact. First, a simple change to the storage representation of the SAC-1 list elements resulted in a 2:1 performance improvement. Second, taking advantage of 64-bit integer arithmetic allowed the use of 31-bit "digits" in the integer arithmetic procedures, reducing the execution time by a factor of 5 compared to the 14-bit "digits" in the published Buchberger algorithm.

As a means of measuring the effect of the changes to the SAC-1 procedures, a series of implementations of Algorithm 628 was generated, each incorporating progressively more of the changes. Execution time was measured for a particular problem [Trinks, 1978; Matzat, 1979] which had been reported as taking 14 minutes on a Univac 1108 and 42 minutes on an IBM 370/155. The nature of the modifications, and their effect on overall computation time, are reported here for the three classes of changes — generic, register-width, and vectorization. The cumulative effect was a reduction of the execution time from 45 minutes (the initial C-120 time) down to 59.8 seconds. On the Convex C-210, a faster member of the C-series, the execution time for the same test case is 23.7 seconds.

Generic changes

The implementation of lists in SAC-1 reflects the time this code was written, when memory was expensive and a user with a quarter of a megabyte considered himself extremely well off. Each list element has four components: a type, a usage count, a tail, and a data field which (depending on the type) may be simple data or a pointer to another list element. To conserve space the type, count, and tail are packed into a single array element. Reference to these fields then requires shifting and masking, which are implemented in standard Fortran by divisions and "mod" operations. This is costly for two reasons: first, division is usually the slowest arithmetic operation on any digital computer, and second, examining or modifying any of these fields requires a function or subroutine call. When memory was scarce, this was probably a "necessary evil"; but today, when even small machines are measured in megabytes of storage, the economics of trading time for space do not make sense. Accordingly the SAC-1 SPACE array was replaced by four integer arrays named TYPE, COUNT, TAIL, and FIRST (corresponding to the functions that existed to extract their contents, which functions were of course eliminated along with the corresponding subroutines STYPE, SCOUNT, SSUCC, and ALTER, used to set new values in the respective fields.)

It is worth noting that, although SAC-1 was designed with portability in mind and almost all machine dependencies are represented by parameters of one sort or another, as published the SAC-1 list processing procedures assume that Fortran integers are at least 32 bits in length. By removing the packed fields in the SPACE array, we now have a version of SAC-1 that can run on 16-bit mini- and microcomputers, an unexpected benefit of the effort to make it run efficiently on supercomputers.

Two functions, PFA and PFL, are called hundreds of thousands of times in the test case. Their function is to append an atom at the head of a list, or to create a list from two given lists. Because they are primitives in SAC-1 it is necessary to keep them as separate Fortran procedures, but the procedure call overhead is significant with respect to the very small amount of computation that they do. The Convex Fortran compiler includes an option for "automatic in-lining" of procedures — in essence, it replaces a "subroutine jump" machine instruction with a copy of the called procedure, just as a literal interpretation of the Algol 60 "copy rule" requires. The PFA and PFL subroutines seemed like good candidates for in-lining.

After these changes — the restructuring of the SPACE array, the concomitant elimination of the procedure calls to access each of the fields of a list element, and the in-lining of PFA and PFL — were implemented, the test case ran in 22 minutes. This is better than a 2:1 reduction from the original time of 45 minutes.

Architectural changes — register width

Also independent of the vector architecture of the computer is the width of its registers. While 64-bit floating point numbers are common, most computers available today are limited to 32 bits for integer computation. The Convex C-series implements 64-bit integer arithmetic in hardware, and provides a set of extensions to Fortran (an INTEGER*8 data type, and intrinsics INT4 and INT8 to convert 8-byte integers to 4-byte and vice versa) to give application programs access to the hardware capability. Since multiple-precision multiplication and division require execution time which is roughly proportional to the square of the number of digits, it stands to reason that reducing the number of digits by a factor of 2 should reduce the execution time by a factor of 4. (The SAC-1 multiplication routine does not take advantage of the esoteric algorithms that are less than order n-squared, and most of these algorithms do not begin to perform at their optimal rates until the length of the numbers is exceedingly large.) The analysis of the greatest common divisor algorithm is not as simple, but it spends most of its time doing simple machine divisions as an approximation to the multiple-precision divisions that would otherwise be needed, and again it is apparent that a wider dividend and divisor should improve its performance.

SAC-1 arithmetic routines represent integers as a list of digits modulo BETA, a Fortran integer whose value is set by the user. The largest positive digit value is BETA - 1; digits may be either positive or negative. Hence BETA must be chosen small enough that the product BETA * (BETA - 1) can be represented as an integer (accommodating a product of two digits plus a third carry digit). It is convenient for other reasons to let BETA be a power of 2.

Algorithm 628 selects 14-bit digits and takes 42 minutes to solve the Trinks problem on an IBM 370/155, whereas Trinks reported solving the problem in one third of the time on a Univac 1108. There is not that much difference between the two processors in terms of arithmetic speed or memory cycle time. However, the 1108 is a 36-bit machine and therefore supports 17-bit digits for multiple-precision integer arithmetic. On the basis of my experience with the Convex C-120, I believe this accounts for most of the difference in the two published times.

Instead of 14 bits, Winkler *et al* could have chosen 15 bits; apparently they did not because as a 16-bit integer the number 32768 will appear to be negative. As a preliminary experiment I re-ran the test program using 15-bit digits. The result was a 15-minute run, compared to 22 minutes with 14-bit digits. Encouraged by this I proceeded to revise the SAC-1 routines to use full 64-bit arithmetic, i.e. using digits of 31 bits each. This required some syntactic changes to the routines in order to get the basic arithmetic to work right (when, for example, it was desired to obtain the 64-bit product of two 32-bit numbers). With these changes the test case ran in 4.5 minutes, a 5:1 reduction from the original 14-bit implementation.

Vectorization changes

Having reduced the test case from 45 minutes to 4.5 minutes, we are at last ready to discuss the changes to the SAC-1 arithmetic routines to take advantage of the vector architecture of the Convex C-series processors. The strategy here involves several observations. First, procedure calls tend to inhibit vectorization because (even if the compiler knows how the called procedures are implemented) they make the dataflow analysis of the program much more complex than it otherwise would be. Second, accessing data in linked lists will confuse the compiler for much the same reason — it has trouble separating things used as pointers from things used as data, and even the best vectorizing compilers will hardly ever be clever enough to create a temporary array for a set of data items that the program references as individual list elements. That is still the programmer's responsibility. Third, a more global view of certain computations may allow improvements such as the accumulation of several additive or multiplicative carries in a single temporary variable before storing a result back into a proper SAC-1 list. These observations are prompted by analysis of the usage of the SAC-1 subroutines ADD3, MPY, and QR, which compute respectively the two-digit sum of three one-digit inputs, the two-digit product of two one-digit factors, and a quotient and remainder (one digit each) from the division of a two-digit dividend by a one-digit divisor. These routines proved to be the primary obstacles to vectorization of SAC-1 arithmetic.

The routines which invoke ADD3, MPY, and QR are ISUM, IDIF, IPROD, IQR, IQRS, and IGCD, the SAC-1 functions for the four basic arithmetic operations and greatest common divisor. (ISUM, although included in the text of Algorithm 628, is never actually used by it; IQRS is a special case of a multiple-precision dividend divided by a one-digit divisor, returning a multiple-precision quotient and a one-digit remainder.) In the case of IPROD, certain sequences of multiplication and addition of partial products could indeed be shortened by taking a more global view of carry accumulation. Further, although QR always produces both a quotient and a remainder, not all calls to QR use both of its outputs, so as the algorithms were modified to eliminate the calls to QR, it was sometimes possible to eliminate some extraneous computations. Finally, division of a positive number by BETA, the SAC-1 integer modulus, can be replaced by the ISHIFT function, a MIL-STD-1753 extension to Fortran which the Convex compiler converts to a binary shift instruction (much quicker than the 64-bit integer divide instruction). This last improvement is of course not possible in QR, which must be general enough to accommodate negative dividends and arbitrary divisors.

These changes reduced the execution time for the test case from 273 seconds (4.5 minutes) to 118 seconds — a 23:1 improvement over the initial implementation, and still the vector registers had not come into play. At this point runtime profiling software indicated unmistakably that by far the largest part of the computation time was being spent in IPROD, IQR, and IGCD, with IGCD the most demanding of all. These, then, became the obvious candidates for vectorization. The strategy was to change them internally to operate on Fortran arrays instead of SAC-1 lists, although their arguments would continue to be lists in order to make the changes transparent to the calling routines.

The published version of IQR calls IPROD to compute the product of the divisor by a trial quotient digit. Since this is a special case of multiplication, and since the intended changes to IQR would mean that an array would have to be copied back into a list before calling IPROD, and the list returned by IPROD would have to be copied back into an array before IQR could continue, replacing the call to IPROD with an integrated multiplication loop seemed advisable. This substantially reduces the number of calls to IPROD; the timings which follow discuss the changes to IQR first, and present the effect of vectorizing IPROD only after the calls from IQR have been eliminated, in order to avoid a misleading appearance that changes to IPROD had more impact than they actually did.

There are two generic SAC-1 routines which call IQR. These are IQ and IREM, which return to the caller only the quotient or only the remainder, discarding the other output of IQR. It turns out that Algorithm 628 makes no calls to IQR directly; all calls to IQR come from IQ or IREM. Further, IGCD is the only source of calls to IREM. In order to avoid copying from arrays to lists and from lists to arrays within IGCD, its calls to IREM were replaced with calls to a new procedure, GCDREM, whose arguments are arrays rather than lists. (This is discussed in greater detail below.) Since this eliminated all calls to IREM, the only source of calls to IQR was now IQ, which would immediately throw away the remainder returned by IQR; this could be avoided by copying the IQR code, except for the part that preserved the remainder, directly into the IQ source module, eliminating one level of subroutine calls and avoiding unnecessary work.

Three loops in IQR were adapted to the Convex vector architecture. Multiplying the divisor by a single digit of the trial quotient and subtracting the product from the remaining part of the dividend, and the rarely-executed correction step when the initial estimate of the quotient digit proves too large, were easy to vectorize. With BETA restricted to powers of 2, adjusting the results modulo BETA reduced to vectorizable shifts and carry-free additions in most cases. The net improvement was a reduction in execution time from 118 down to 108 seconds, a modest 10%, of which 6% was due to the changes in the program logic (i.e., the 6% was realized without invoking the vector register level of compiler optimization) and 4% was due to the vector hardware.

Adapting IPROD to the vector architecture gave similarly modest improvement. Forming partial products vectorized well, as expected, and some of the carry propagation could be vectorized. A final stage of carry propagation, however, had to be left in "scalar" mode because input to each step is dependent on the output of the preceding step. (Working in assembly language, Melenk and Neun at this workshop showed a technique by which all carry steps could be vectorized, using the "population count" hardware instruction to determine when no recursive carries remained. This technique would be applicable to the Convex C-series as well, but it is not accessible through Fortran.) The net effect was to reduce the execution time down to 102 seconds, a 6% improvement of which only 2% was directly attributable to the vector registers.

Changes to IGCD made a more substantial difference in performance, although the effect was still small compared to the impact of 64-bit integer arithmetic and the changes to the memory representation of the SAC-1 list elements. The SAC-1 implementation of IGCD is very similar to the algo-rithmic treatment by Knuth [1981, pp. 316-339, 593-599]. Since I had already decided to limit BETA to powers of 2, I was able to implement IGCD with a binary algorithm, a method recommended by Knuth which would be difficult for arbitrary choices of BETA. Also, a special terminating sequence was coded for the case where the two operands had both been reduced to two SAC-1 elements, since any 2-digit SAC-1 quantity is by definition representable as a Fortran integer. The most significant algorithmic change, however, was to replace the calls to IREM, by which IGCD would reduce the "dividend" modulo the "divisor", with calls to a newly created routine, GCDREM, which unlike IREM operated on arrays rather than lists. This eliminated a great deal of memory activity (copying data between lists and arrays) and GCDREM, like IQR on which it was modeled, vectorized well. Other important functions within IGCD that vectorized well involved scaling the operands before the trial divisions of the "Lehmer procedure" and computing new residues for the next stage of the problem.

These modifications to IGCD reduced the total execution of the Groebner basis program, for the "Trinks-Matzat" problem, to 59.8 seconds, about a 42% improvement of which half was attributable to algorithmic changes and half to the vector hardware. This emotionally satisfying result, running the test case in less than one minute, is a better than 45:1 improvement over the original implementation.

Before concluding, it is necessary to report one unsuccessful attempt at further vectorization of the greatest common divisor algorithm. The bulk of the time is spent in a procedure first proposed by D. H. Lehmer, in which multiple-precision divisions are replaced by single-precision approximations. In Lehmer's procedure, two single-precision divisions are performed at each step, such that one of them produces a quotient which is slightly too small, and the other a quotient which is slightly too large. As long as the integer representation of the two quotients is the same, these two divisions represent the same integer quotient that would be obtained by multiple-precision divisions. The Lehmer procedure continues single-precision division and remaindering until the two integer quotients are unequal, at which time it performs in multiple precision the cumulative multiplications and subtractions that would have been done by a naive multiple-precision algorithm. Then the program may return to the Lehmer procedure, or a single multiple-precision division may be required.

This process involves two independent divisions, and as long as the quotients are equal, four independent multiply-add sequences. These could be implemented as vector operations with vector lengths of 2 and 4 respectively. However, there is a time penalty associated with starting up vector operations, and when this experiment was tried the execution time was actually a few seconds longer than in the non-vectorized case. Convex users generally report that vector lengths of 4 or less are too short to justify using the vector registers, and this experiment certainly supported that conclusion.

Summary and conclusions

Although the vector architecture of the Convex C-series was the motivation for the changes to the SAC-1 procedures, memory and register width proved to have an even greater impact on overall performance. Analysis of the rather disappointing results of vectorization showed that most of the vectors were of length 10 or less, still borderline as far as efficiency is concerned. It is probable that with substantially larger problems, leading to vector lengths of 50 or 100 or more, vectorization of IQR, IPROD, and IGCD will prove to be more significant.

This presentation of the changes to the SAC-1 routines is actually somewhat idealized. They were originally implemented in a much more random fashion, trying things as they occurred to me. However, to determine the actual effect of each of the changes after having once reduced the execution time below one minute, I went back to the original version of the code and then re-introduced the changes systematically in the order they are presented here. Thus I believe the relative improvements shown here reflect the real world quite faithfully.

There is further research to be done with respect to this exercise. First, the vectorizations of IQR, IPROD, and IGCD have introduced an implicit upper limit on the length of a SAC-1 integer, based on the Fortran array declarations that define the working storage for these routines. This limitation can be eliminated, by implementing each of these routines at two levels: when the length of the inputs is less than the Fortran array lengths, use the optimized versions, otherwise consider the inputs to consist of "hyper-digits" each of which can be manipulated through the vector logic. This will entail reviving the ADD3, MPY, and QR routines, this time operating on very large quantities and calling IPROD and IQR (or renamed equivalents to avoid recursion in standard Fortran) to perform vector multiplications and divisions. A "multi-modular" representation, described by Melenk and Neun at this workshop, has some very attractive features and can be expected, based on their experience on the Cray-1, to achieve a further improvement by a factor of 5 or more. It is somewhat of an emotional issue, to decide whether this could still be considered an implementation of SAC-1 arithmetic, or whether it would have to be considered an alternative to SAC-1. Finally, the high end of the Convex product line includes systems with two to four closely coupled parallel processors, supporting an MIMD computational model in addition to the SIMD model of the vector registers. I hope soon to investigate whether any part of the SAC-1 arithmetic routines could be further accelerated by fine-grained parallelism on these processors.

References

Collins, G. E. 1971a. The SAC-1 list processing system. Tech. Rep. 129, Computer Science Dept., Univ. of Wisconsin, Madison.

Collins, G. E. 1971b. The SAC-1 integer arithmetic system. Tech Rep. 156, Computer Science Dept., Univ. of Wisconsin, Madison.

Knuth, D. E. 1981. The art of computer programming: seminumerical algorithms. 2nd Edition. Reading, Mass: Addison Wesley.

Matzat, B. H. 1979. Konstruktion von Zahlkörpern mit der Galoisgruppe M_{11} über $\mathbb{Q}(\sqrt{-11})$. Manuscripta Mathematica 27: 103-111.

MIL-STD-1753. Military Standard: FORTRAN, DOD supplement to American National Standard X3.9-1978. Hq, U. S. Air Force/ACDX, Washington, DC 20330, 9 November 1978.

Trinks, W. 1978. Über B. Buchbergers Verfahren, Systeme algebraischer Gleichungen zu lösen. Jour. Number Theory 10: 475-488.

Winkler, F., B. Buchberger, F. Lichtenberger, and H. Rolletschek. 1985. Algorithm 628: an algorithm for constructing canonical bases of polynomial ideals. ACM Trans. Math. Software 11(1): 66-78 (full source code on microfiche supplement 93).

addition to the SIMD model of the vector registers. I hope soon to investigate whether any part of the SAC-I arithmetic routines could be further accelerated by fine-grained parallelism on these processors.

References

Collins, G. E. 1971a. The SAC-1 list processing system. Tech. Rep. 129, Computer Science Dept., Univ. of Wisconsin–Madison.

Collins, G. E. 1971b. The SAC-1 integer arithmetic system. Tech. Rep. 156, Computer Science Dept., Univ. of Wisconsin, Madison.

Knuth, D. E. 1981. The art of computer programming: seminumerical algorithms. 2nd Edition. Reading, Mass.: Addison Wesley.

Matzat, B. H. 1979. Konstruktion von Zahlkörpern mit der Galoisgruppe M_{11} über $Q(\sqrt{-11})$. Manuscripta Mathematica 27, 103-114.

Mil-STD-1753. Military Standard: FORTRAN, DOD supplement to American National Standard X3.9-1978. Hq. U. S. Air Force AF/ADX, Washington, DC 20330, 9 November 1978.

Trinks, W. 1978. Über B. Buchbergers Verfahren, Systeme algebraischer Gleichungen zu lösen. Jour. Number Theory 10, 475-488.

Winkler, F., B. Buchberger, F. Lichtenberger, and H. Rolletschek. 1985. Algorithm 628: an algorithm for constructing canonical bases of polynomial ideals. ACM Trans. Math. Software 11(1): 66-78 (full source code on microfiche supplement 93).

Some Issues in Designing Algebraic Algorithms for the CRAY X-MP

Jeremy R. Johnson
Department of Computer and Information Science
The Ohio State University
Columbus, OH 43210

1 Introduction

This paper is an abstraction of the work done in the author's master's thesis [4], where more details can be found. This work began as a study of the implementation of the SAC-2 computer algebra system [3] on the CRAY X-MP supercomputer [1]. After some preliminary experiments, it became apparent that the raw computing power of the X-MP could be useful for algebraic computation; however, the basic architectural features of the X-MP were not being used. The design of the SAC-2 algorithms and their implementation prevents the X-MP from using vector instructions to obtain the speed-up that might be possible. Some of these obstacles are due to the mismatch of the architecture of the X-MP and the requirements of computer algebra algorithms; however, by modifying some of the algorithms used in SAC-2 and by restricting the problems we intend to solve, it is possible to take advantage of the architecture of the X-MP and similar vector machines.

In this paper we present some techniques for modifying algebraic algorithms to fit the architecture of the X-MP. By modifying an algorithm to fit the architecture of a machine, we mean altering the algorithm's data structures and data flow, so that the algorithm runs efficiently on that machine. For the X-MP, we have tried to do this by using vector instructions, minimizing memory operations, keeping vectors in the

COMPUTER ALGEBRA AND PARALLELISM ISBN 0-12-221100-6

vector registers as long as possible, and segmenting the operations on vectors so that overlap can take place. For the algorithms we have considered, there are some mathematical properties of the computations that can be used for this modification. Some of the techniques we have used are decomposition, reordering, and change of representation. In many cases, these techniques arise naturally by considering the algebraic structures involved.

2 CRAY Architecture and Programming Techniques

Before presenting an example of modifying an algorithm for polynomial multiplication to utilize X-MP architecture, we need to review some of the important features of the X-MP. The X-MP is a highly pipelined machine designed to execute vector instructions. A vector instruction takes elements from a vector register, pipes them through the appropriate functional unit, and sends the results to another vector register. Each of the functional units, for example floating point addition and floating point multiplication, is pipelined so that once the first result is produced, a new result is produced every clock period. For example, two vector registers, containing 64 elements each can be added in 69 clock periods (CPs). This is the time needed to produce the first result (6 CPs) plus one CP each for the remaining elements. If this were done with a loop of scalar adds 384 CPs would be required.

Since all vector instructions operate on vectors contained in vector registers and each vector register contains at most 64 elements, it is necessary to partition large vectors into segments containing 64 elements or less. Each segment must be loaded into a vector register, where the appropriate operation can be performed, and then stored back to memory. We call a sequence of such operations that carry out a vector instruction on a large vector a supervector instruction, to distinguish it from the hardware supported vector instructions. A supervector instruction is implemented in software in a manner very similar to the way the built-in vector instructions are carried out in hardware. In fact, we can think of the X-MP as a programmable pipeline, where the stages of the pipeline are carried out in the functional units (we consider memory as a functional unit) and temporary results are stored in the vector registers. This model is reasonable, since the functional units are independent, which implies that vector instructions using separate functional units and vector registers can execute concurrently.

For example, suppose we wish to add two supervectors. In this case, we use a loop that implements a pipeline with three stages. The first stage loads a segment from each supervector into a vector register, the second stage adds these segments, and the final stage stores the result back to memory. The benefit of segmenting this computation into these three stages is that the operations performed in each stage can be overlapped. That is, while a pair of segments of the supervector is being loaded, another two can be added, and the result of the previous addition can be stored. This overlap can be obtained on the X-MP, since the memory operations can take place simultaneously with the additions. It is also essential that the X-MP has three separate memory ports, two for loading vector registers and one for storing a vector register, otherwise the memory operations could not be effectively overlapped.

We now give an example of modifying an algorithm to obtain vector instructions and proper segmentation. The example we give is matrix-vector multiplication, which will be important when we consider polynomial multiplication. If we let

$$A = \begin{pmatrix} a_{11} & \cdots & a_{1n} \\ \vdots & \ddots & \vdots \\ a_{n1} & \cdots & a_{nn} \end{pmatrix} \text{ and } x = \begin{pmatrix} x_1 \\ \vdots \\ x_n \end{pmatrix},$$

then the computation can be written as Ax.

Rather than perform this operation as n inner products, as is usually done on non-vector machines, we utilize the well known transformation:

$$Ax = x_1 A_{*1} + \cdots + x_n A_{*n}$$

where A_{*j} is the j-th column of A. This can be obtained by writing A as a sum of matrices containing the columns of A and distributing x over this sum. To perform this computation, we multiply a column of the matrix by an element of the input vector and add this vector to a vector containing the partial sum corresponding to the previous columns. The benefit of this reorganization is that we replace inner product computations, which are hard to vectorize, with a series of "saxpy" transformations (scalar a times a vector x plus a vector y) that can be carried out on the X-MP. Furthermore, the operations can be segmented so that vector loads, scalar-vector multiplies, vector adds, and vector stores can be overlapped. If n is larger than 64, the matrix should be partitioned into bands of 64 elements or less. Another important benefit of segmenting the computation into operations with 64 elements or less is that the partial sum can be accumulated in the vector registers. This saves many memory operations, since partial results do not have to be stored and reloaded.

3 Design and Modification of an Algorithm for Polynomial Multiplication

In this section we derive an algorithm for multiplying polynomials with integer coefficients. This algorithm uses vector instructions and obtains effective segmentation. In general, multiplying polynomials with arbitrarily large integer coefficients is not well suited to the X-MP, since the sizes of the operands involved are not uniform. This lack of uniformity prevents the coefficients vectors from fitting in the vector registers, and hence the coefficient operations can not be performed with vector instructions. We will use modularization and reconstruction via the Chinese Remainder Theorem (CRT) to make the problem more uniform and obtain vectorization. The CRT will also be used to obtain segmentation.

For our purposes, the CRT is best thought of as a theorem dealing with different representations of a ring. In particular for the ring of polynomials $Z[x]$, the CRT decomposes $Z[x]$ with respect to a set of relatively prime polynomials. If $f_1, \ldots, f_n$ are pairwise relatively prime elements of $Z[x]$, we get the following isomorphism into a direct product of simpler rings:

$$Z[x]/\prod f_i \cong Z[x]/f_1 \times \cdots \times Z[x]/f_n.$$

Moreover, the CRT gives an orthogonal set of idempotents $e_1, \ldots, e_n$, i.e.,

1. $e_i e_j = 0$ if $i \neq j$.
2. $e_i e_i = 1$.
3. $e_1 + \cdots + e_n = 1$.
4. $e_i \equiv 0 \pmod{f_j}$ if $i \neq j$ and $e_i \equiv 1 \pmod{f_i}$.

These idempotents can be used to make the preceding isomorphism explicit. If $A \in Z[x]$ projects onto $(A_1, \ldots, A_n)$ where $A_i \in Z[x]/f_i$, then $A = A_1 e_1 + \cdots + A_n e_n$. We will give an example later that will clarify these concepts.

For polynomial operations in $Z[x]$, there exists a coefficient bound N and a degree bound d such that operations in $Z[x]$ are isomorphic to operations in $Z_N[x]/F(x)$, where $F(x)$ is a polynomial of degree d. In order to design effective polynomial algorithms for the X-MP, we will use the CRT in the ring $Z_N[x]/F(x)$ to find an appropriate representation of polynomials for vectorization and segmentation. First of all, we obtain a uniform representation of polynomials by mapping the given polynomial in $Z[x]$ onto $Z_p[x]$, where p is a prime that fits in a single word of the machine being used. Uniformity is obtained since each of the coefficients will fit in a single word. The CRT is used to recover the desired integer result from the computations in $Z_p[x]$. If we choose $N = p_1 \cdots p_t$, a product of t distinct wordsized primes, the CRT gives the following isomorphism:

$$Z_N[x] \cong Z_{p_1}[x] \times \cdots \times Z_{p_t}[x].$$

This isomorphism allows us to perform polynomial operations in $Z_N[x]$ with the uniform operations in $Z_p[x]$.

Multiplication in each component $Z_p[x]$ of this direct product can be performed with vector instructions (see [4]); however, if the degree of the product is greater than 63, extra memory operations will be required due to improper segmentation. The CRT can be used with polynomials of the form $x^{64} - \alpha$ to obtain better segmentation. Polynomial operations in $Z_p[x]/x^{64} - \alpha$ are perfectly segmented and can be carried out entirely in the vector registers. Here, the CRT is used with several polynomials of the form $x^{64} - \alpha$. Since the $\gcd(x^{64} - \alpha_i, x^{64} - \alpha_j) = 1$ when $\alpha_i \neq \alpha_j$, the CRT can be applied to give the following isomorphism provided the α_i's are distinct:

$$Z_p[x]/\prod_{i=1}^{\tau}(x^{64} - \alpha_i) \cong Z_p[x]/(x^{64} - \alpha_1) \times \cdots \times Z_p[x]/(x^{64} - \alpha_\tau).$$

This isomorphism segments polynomial operations by performing them in $Z_p[x]/x^{64} - \alpha$. If enough factors $x^{64} - \alpha_i$ are used so that 64τ, the degree of $F(x) = \prod_{i=1}^{\tau}(x^{64} - \alpha_i)$, is greater than the degree bound d, then integral polynomial operations can be recovered from this representation.

In order to better understand polynomial operations in $Z_p[x]/x^n - \alpha$, we need to introduce the regular representation of this algebra. Polynomials can be thought of as vectors of coefficients with respect to the basis $\{1, x, \ldots x^{n-1}\}$. The regular representation is the mapping ρ from $Z_p[x]/x^n - \alpha$ into $n \times n$ matrices such that $\rho(A)B = AB$ for the polynomials A and B. To get a picture of this matrix observe that

$$\rho(x) = \begin{pmatrix} 0 & \cdots & 0 & \alpha \\ 1 & \ddots & & 0 \\ & & \ddots & \\ 0 & & 1 & 0 \end{pmatrix} = C$$

with respect to the basis $\{1, x, \ldots, x^{n-1}\}$ since $x \cdot 1 = x, x \cdot x = x^2, \ldots, x \cdot x^{n-1} = x^n = \alpha$. This is the companion matrix of $x^n - \alpha$. Since ρ is linear, an arbitrary polynomial $A(x) = a_{n-1}x^{n-1} + \cdots + a_1 x + a_0$, gets mapped to $\sum_{i=0}^{n-1} a_i C^i$. Since the minimal polynomial of C is $x^n - \alpha$, ρ is an isomorphism and the representation is faithful. For example, if $n = 4$ the polynomial $a_3x^3 + a_2x^2 + a_1x + a_0$ gets mapped to

$$\begin{pmatrix} a_0 & \alpha a_3 & \alpha a_2 & \alpha a_1 \\ a_1 & a_0 & \alpha a_3 & \alpha a_2 \\ a_2 & a_1 & a_0 & \alpha a_3 \\ a_3 & a_2 & a_3 & a_0 \end{pmatrix}.$$

We can now use this representation to design an algorithm for multiplying two polynomials A and B in $Z_p[x]/x^n - \alpha$. We simply use the matrix-vector multiplication described in section 2. In this case, the matrix is $\rho(A)$ and the input vector is the coefficient vector corresponding to B. If we let

$$A = \begin{pmatrix} a_0 \\ a_1 \\ \vdots \\ a_{n-1} \end{pmatrix}, B = \begin{pmatrix} b_0 \\ b_1 \\ \vdots \\ b_{n-1} \end{pmatrix},$$

then the computation can be written as

$$b_0 A + b_1 \rho(x) A + \cdots + b_{n-1} \rho^{n-1}(x) A.$$

$\rho(x)$ rotates the coefficient vector corresponding to A and multiplies the last coefficient by α. For example,

$$\begin{pmatrix} 0 & 0 & 0 & \alpha \\ 1 & 0 & 0 & 0 \\ 0 & 1 & 0 & 0 \\ 0 & 0 & 1 & 0 \end{pmatrix} \begin{pmatrix} a_0 \\ a_1 \\ a_2 \\ a_3 \end{pmatrix} = \begin{pmatrix} \alpha a_3 \\ a_0 \\ a_1 \\ a_2 \end{pmatrix}.$$

We can use this observation to prevent storing all the columns of the matrix. In fact only the first column, which contains the coefficients of the polynomial, needs to be stored. As the computation proceeds, instead of loading the columns, the elements of the vector register containing the previous column are rotated, and the appropriate element is multiplied by α. Since n has been chosen to equal 64, all of the coefficients fit in a single vector register, and since rotation can be performed without using memory operations, the entire matrix operation can be done in the registers. This fact combined with the general properties of matrix-vector multiplication discussed before, leads to an algorithm adapted to the X-MP.

In general, if we are doing arithmetic in $Z_p[x]/x^{64}-\alpha_1 \times \cdots \times Z_p[x]/x^{64}-\alpha_r$, we will have the computation corresponding to the direct sum

$$\begin{pmatrix} \rho_{\alpha_1}(A) & & 0 \\ & \ddots & \\ 0 & & \rho_{\alpha_r}(A) \end{pmatrix},$$

where $\rho_{\alpha_i}(A)$ is the 64×64 matrix representation of A in $Z_p[x]/x^{64}-\alpha_i$. Therefore the entire operation is segmented for the X-MP.

In order to complete the algorithm, we need to describe how to convert into and out of this representation. Converting polynomials into this representation involves projecting the polynomial onto the various factors in the isomorphism given by the CRT. If we let $N = p_1 \cdots p_t$ and $F(x) = (x^{64} - \alpha_1)\cdots(x^{64} - \alpha_r)$, the change of representation is summarized by the following two isomorphisms

$$\begin{aligned} Z_N[x]/F(x) &\cong Z_{p_1}[x]/F(x) \times \cdots \times Z_{p_t}[x]/F(x) \\ &\cong Z_{p_1}[x]/(x^{64} - \alpha_1) \times \cdots \times Z_{p_1}[x]/(x^{64} - \alpha_r) \\ &\quad \times \cdots \times \\ &\quad Z_{p_t}[x]/(x^{64} - \alpha_1) \times \cdots \times Z_{p_t}[x]/(x^{64} - \alpha_r). \end{aligned}$$

Changing into this representation requires two steps. The first step projects the coefficients onto Z_p for each of the primes, thus limiting the size of the coefficients. The second step projects polynomials in $Z_p[x]/F(x)$ onto the factors $Z_p[x]/x^{64} - \alpha_i$ which reduces the degree of the polynomials to 63 or less. The inverse operation, which lifts polynomials in the various factors back to polynomials in $Z_N[x]/F(x)$, is performed with the idempotents given by the CRT. We will consider the operations involved in the second step in some detail, while deferring the other step to the example at the end of this section.

We now discuss how to perform the projection and lifting operations needed to convert from polynomials in $Z_p[x]/(x^{64} - \alpha_1)\cdots(x^{64} - \alpha_r)$ to polynomials in $Z_p[x]/(x^{64} - \alpha_1) \times \cdots \times Z_p[x]/(x^{64} - \alpha_r)$ and vice versa. The emphasis of our discussion will concern vectorization and segmentation of these operations. Before discussing this we look at projection onto $Z_p[x]/(x-\alpha)$, which is just evaluation at the point α. For example, if $A(x) = a_m x^m + \cdots + a_1 x + a_0$, projection onto $Z_p[x]/(x-\alpha)$ can be written as the dot product computation

$$(1\ \alpha\ \cdots\ \alpha^m)\begin{pmatrix} a_0 \\ a_1 \\ \vdots \\ a_m \end{pmatrix}.$$

This computation is not well suited to the X-MP, since it can not be performed easily with vector instructions. However, if we needed to evaluate $A(x)$ at several points we could obtain a vectorized algorithm. To see this, observe that the computation can be written as a square, when $\tau = n+1$, Vandermonde matrix times a vector:

$$\begin{pmatrix} 1 & \alpha_1 & \cdots & \alpha_1^n \\ 1 & \alpha_2 & \cdots & \alpha_2^n \\ \vdots & \vdots & \cdots & \vdots \\ 1 & \alpha_{n+1} & \cdots & \alpha_{n+1}^n \end{pmatrix}\begin{pmatrix} a_0 \\ a_1 \\ \vdots \\ a_n \end{pmatrix}.$$

The algorithm given in section 2 for matrix-vector multiplication can be thought of as replacing a scalar evaluation algorithm with a vector algorithm if we simply replace scalar operations with vector operations:

$$a_0\begin{pmatrix} 1 \\ \vdots \\ 1 \end{pmatrix} + a_1\begin{pmatrix} \alpha_1 \\ \vdots \\ \alpha_{n+1} \end{pmatrix} + \cdots + a_n\begin{pmatrix} \alpha_1^n \\ \vdots \\ \alpha_{n+1}^n \end{pmatrix}.$$

Clearly Horner's method could be used as in the scalar case.

An alternative way of obtaining a vector algorithm would be to replace $x-\alpha$ by $x^n-\alpha$. To see why this leads to a vector operation, we partition the polynomial $A(x)$ into segments of n terms each:

$$A(x) = A_m(x)x^{mn} + \cdots + A_1(x)x^n + A_0(x),$$

where each $A_i(x)$ is a polynomial of degree at most $n-1$. Then projection onto $Z_p[x]/x^n-\alpha$ amounts to replacing x^n by α. Thus we are left with the polynomial

$$A(x) = A_m(x)\alpha^m + \cdots + A_1(x)\alpha + A_0(x).$$

This operation multiplies each segment $A_i(x)$ by α^i, and then adds all of the segments together. If we write this as a matrix operation on the coefficient vector corresponding to A, we get a block dot product computation on the segments of coefficients corresponding to $A_i(x)$:

$$(I_n \ \alpha I_n \ \cdots \ \alpha^m I_n) \begin{pmatrix} A_0 \\ A_1 \\ \vdots \\ A_m \end{pmatrix}.$$

This block computation can described with tensor product notation. The tensor product of two matrices, replaces the ij-th entry of the first matrix, with the block equal to the ij-th entry times the second matrix. For example, if we let $\otimes$ denote the tensor product, the block dot product computation can be represented by

$$(1 \ \alpha \ \cdots \ \alpha^m) \otimes I_n \begin{pmatrix} A_0 \\ A_1 \\ \vdots \\ A_m \end{pmatrix}.$$

The way to interpret this operation is as evaluation on vectors of length n. If n is chosen to be 64, this operation would be segmented for the X-MP. In terms of operations for the X-MP, this is nothing more than a special case of a supervector dot product.

In our application of the CRT, we will be projecting onto several factors of the form $Z_p[x]/x^{64} - \alpha$. Using tensor product notation this amounts to the matrix operation $(V \otimes I_{64})A$, where V is a Vandermonde matrix and A is the coefficient vector. Each row of the Vandermonde matrix corresponds to projecting onto a different factor $Z_p[x]/x^{64} - \alpha$.

The benefit of using this matrix notation is that it gives us an easy grasp of the inverse operation that lifts the factors $Z_p[x]/x^{64} - \alpha$ back to polynomials in $Z_p[x]/\prod(x^{64}-\alpha_i)$. This inverse operation corresponds to the inverse of the matrix operation $V \otimes I_{64}$.

Before trying to understand this vector operation, we again look at the case where each factor corresponds to a linear polynomial $x-\alpha$. In this case, we know V is invertible if there are $n+1$ distinct points of evaluation α_i. This is another way of saying that a polynomial of degree n is determined if we know its value at $n+1$ distinct points. In fact, the inverse of V corresponds to Langrangian interpolation. The matrix V^{-1} can be written down from the idempotents given by the CRT. Furthermore, since the idempotents can be explicitly constructed, V^{-1} can easily be computed [4] and [5].

We now return to the lifting operation given by $(V \otimes I_{64})^{-1}$. Since the inverse of a tensor product of matrices is the tensor product of the inverses (provided the matrix is invertible), $(V \otimes I_{64})^{-1} = V^{-1} \otimes I_{64}$. Thus the lifting operation is a vector operation similar to the corresponding projection operation. It can be thought of as interpolation on vectors of length 64.

We are now ready to summarize the complete polynomial multiplication algorithm. It is given by the following three steps.

1. Projection onto $Z_p[x]/(x^{64}-\alpha_1) \times \cdots \times Z_p[x]/(x^{64}-\alpha_r)$ for several word-sized primes p. This is done with supervector dot products.

2. Block diagonal matrix-vector multiplication corresponding to multiplication in $Z_p[x]/(x^{64}-\alpha)$.

3. Interpolation using supervector dot products corresponding to the inverse operation of (1).

Since each of these steps can be performed with properly segmented vector instructions, we have an algorithm well suited to the X-MP. We now proceed to give an example of the use of this algorithm.

In our example, we will assume that the vector registers contain only four elements and that the wordsize is large enough to contain one decimal digit. Thus we will use the CRT with single digit primes and polynomials of the form $x^4 - \alpha$. Furthermore, the polynomials we will consider will have degrees less than eight, so that we will only need two factors of the form $Z_p[x]/x^4 - \alpha$. We will use 1 and 2 for the two values of α. Since these two numbers are distinct, the CRT will give us the desired isomorphism. This situation is summarized with the following isomorphisms.

$$\begin{aligned} Z_{15}[x]/(x^4-1)(x^4-2) &\cong Z_3[x]/(x^4-1)(x^4-2) \times Z_5[x]/(x^4-1)(x^4 \\ &\cong (Z_3[x]/(x^4-1) \times Z_3[x]/(x^4-2)) \\ &\times\ (Z_5[x]/(x^4-1) \times Z_5[x]/(x^4-2)). \end{aligned}$$

We choose $A(x) = x^7 + 4x^6 + 13x^5 + 12x^4 + 7x^3 + 11x^2 + 10x + 14$ and $B(x) = 12x^7 + 14x^6 + 2x^5 + x^4 + 9x^3 + 8x^2 + 12x + 4$, as the two polynomials we wish to multiply. Since the coefficients and degrees of the product AB are not bounded by 15 and 8 we can not recover the product as an integer polynomial. If we desired the integral polynomial product we would have to choose larger bounds, and hence more factors in the CRT. However, for our example, we are only interested in showing the various vector operations that are performed.

The algorithm proceeds by projecting A and B onto $Z_3[x]$ and $Z_5[x]$. If we let A_3 denote the component of A in $Z_3[x]$ and A_5 the component of in $Z_5[x]$ we get the pair of polynomials $(A_3, A_5) = (x^7 + x^6 + x^5 + x^3 + 2x^2 + x + 2, x^7 + 4x^6 + 3x^5 + 2x^4 + 2x^3 + x^2 + 4)$. Likewise, we get $(B_3, B_5) = (2x^6 + 2x^5 + x^4 + 2x^2 + 1, 2x^7 + 4x^6 + 2x^5 + x^4 + 4x^3 + 3x^2 + 2x + 4)$. At this point, we observe that the idempotents for $Z_{15}[x]$ are $e_3 = 10$ and $e_5 = 6$. The fact that these are the idempotents can easily be verified. For a small example like this, it is easy to use property (4) of the idempotents to write them down. In larger examples, the extended Euclidean algorithm can be used to construct the idempotents [5]. These idempotents can be used to lift polynomials given by their components in $Z_3[x]$ and $Z_5[x]$ back to $Z_{15}[x]$. For example, $A = A_3e_3 + A_5e_5$ and $B = B_3e_3 + B_5e_5$. Using the properties of the idempotents one can also observe that $AB = A_3B_3e_3 + A_5B_5e_5$. This corresponds to the direct product given by the isomorphism above.

In the next step, we need to project onto $Z_p[x]/(x^4 - \alpha)$, for p equal to

3 and 5 and α equal to 1 and 2. We will do this calculation using the matrix operations discussed previously. For example, we can represent $A_3(x)$ by its vector of coefficients, then projection onto $Z_3[x]/(x^4 - 1) \times Z_3[x]/(x^4 - 2)$ is given by

$$\begin{pmatrix} 1 & 1 \\ 1 & 2 \end{pmatrix} \otimes I_4 \begin{pmatrix} 2 \\ 1 \\ 2 \\ 1 \\ 0 \\ 1 \\ 1 \\ 1 \end{pmatrix} = \begin{pmatrix} 2 \\ 2 \\ 0 \\ 2 \\ 2 \\ 0 \\ 1 \\ 0 \end{pmatrix}.$$

The two by two matrix is the Vandermonde matrix corresponding to the points 1 and 2, and the given tensor product operation evaluates this matrix on vectors of length 4. The fact that 4 was chosen as the degree of the factors onto which we are projecting implies that the operation fits in our hypothetical vector registers of four elements. The vector instructions that carry out this operation are

1. load the first 4 coefficients of A_3 into vector register $V0$.
2. load the second 4 coefficients of A_3 into $V1$.
3. $V2 \leftarrow V0 + V1$.
4. $V3 \leftarrow 2 * V1$.
5. $V4 \leftarrow V0 + V3$.
6. store $V2$ and $V4$.

The resulting vector of the matrix computation should be interpreted as the pair of polynomials $(2x^3 + 2x + 2, x^2 + 2)$ whose coefficients would be stored in $V2$ and $V4$ respectively. These are the components of A_3 in $Z_3[x]/x^4 - 1$ and $Z_3[x]/x^4 - 2$. We denote these components by $(A_{3,1}, A_{3,2})$. Similarly we obtain the components $(B_{3,1}, B_{3,2}) = (x^2 + 2x + 2, x)$ for B_3.

Multiplication of A_3 and B_3 in this representation is given by the matrix direct sum corresponding to the regular representations:

$$\begin{pmatrix} 2 & 2 & 0 & 2 & & & & \\ 2 & 2 & 2 & 0 & & & & \\ 0 & 2 & 2 & 2 & & & & \\ 2 & 0 & 2 & 2 & & & & \\ & & & & 2 & 0 & 2 & 0 \\ & & & & 0 & 2 & 0 & 2 \\ & & & & 1 & 0 & 2 & 1 \\ & & & & 0 & 1 & 0 & 2 \end{pmatrix} \begin{pmatrix} 2 \\ 2 \\ 1 \\ 0 \\ 0 \\ 1 \\ 0 \\ 0 \end{pmatrix} = \begin{pmatrix} 2 \\ 1 \\ 0 \\ 0 \\ 0 \\ 2 \\ 0 \\ 1 \end{pmatrix}.$$

The resulting vector corresponds to components of the product $(A_{3,1}B_{3,1}, A_{3,2}B_{3,2}) = (x+2, x^3+2x)$.

Again, one should observe, that the matrix operations are perfectly segmented to our hypothetical vector machine.

To lift these components back to $Z_3[x]/(x^4-1)(x^4-2)$ we need to know the idempotents. The idempotents are $E_1 = \frac{x^4-2}{1-2} = 2x^4 - 1$ and $E_2 = \frac{x^4-1}{2-1} = x^4 - 1$, since $E_1 \equiv 1 \pmod{x^4-1}$ and $E_1 \equiv 0 \pmod{x^4-2}$ and $E_2 \equiv 0 \pmod{x^4-1}$ and $E_2 \equiv 1 \pmod{x^4-2}$. To interpolate we need to compute $A_{3,1}B_{3,1}E_1 + A_{3,2}B_{3,2}E_2$. This can be represented as the following matrix computation:

$$\begin{pmatrix} -1 & 2 \\ 2 & 1 \end{pmatrix} \otimes I_4 \begin{pmatrix} A_{3,1}B_{3,1} \\ A_{3,2}B_{3,2} \end{pmatrix}.$$

The two by two matrix here is just the inverse of the previous two by two Vandermonde matrix. Observe how the idempotents appear in this matrix. As with evaluation, this tensor product operation implies a vector operation on vectors of length 4. Evaluating the preceding matrix operation we get $x^7+x^5+x^4+2x^3+1$ as the resulting polynomial. One can verify this directly by multiplying A_3B_3. In exactly the same way, one can compute $A_5B_5 = x^7 + x^5 + x^3 + 2x^2 + 4x$, which can be lifted along with A_3B_3 to get the final answer by computing $10A_3B_3 + 6A_5B_5 = x^7 + x^5 + 10x^4 + 11x^3 + 12x^2 + 9x + 10$.

4 Summary

The two key subalgorithms needed for polynomial multiplication, namely supervector dot product and matrix-vector multiplication have been implemented on the X-MP model 48, in the CRAY assembly language. The product of a 64×64 matrix and a 64 element vector of floating point numbers was computed in 0.0405 ms at a utilization rate of 197.34 MFLOPS (millions of floating point operations per second). The supervector dot product of two vectors of length $64 \cdot 2^{10}$ was computed in 0.654 ms at a rate of 200.42 MFLOPS. From these timings we can estimate the time to project, multiply, and lift two polynomials of degree 64000 in $Z_p[x]$. This time is approximately 1300 ms. To see how this estimate was calculated and the programs used to obtain the timings see [4].

It is important to point out the preceding times are for floating point operations instead of integer operations. Since the X-MP does not have a good set of integer instructions, our estimates are somewhat optimistic. Nonetheless, they do give an idea of what vectorization and segmentation can do for polynomial operations on the X-MP.

There are many other problems that need to be addressed if these ideas are to be of practical value in computer algebra. Besides support for operations in Z, one needs to incorporate these algorithms and their data structures into computer algebra systems for machines like the X-MP. Some of these issues have been dealt with in [2], where the authors deal with list processing on the CRAY. This is especially important in computer algebra systems like SAC-2, where lists are a prevalent data structure. Also one needs to find time consuming problems with the need for large polynomial operations that are not too sparse. Finally, one would like to be able to determine in advance, particular problems, or parts of problems suited to the operations that can be performed efficiently on the X-MP. Along these lines, one would also like to do some precomputation, for example, computing idempotents, before letting the CRAY grind away.

The ideas discussed demonstrate how to vectorize and segment algorithms to run efficiently on a vector architecture. Further study is needed to determine if these ideas can be combined with methods for overcoming the problems mentioned above. Finally, with the efficiency gain that the CRAY can offer, problems should be examined to see if the are amenable to the techniques discussed in this paper. Furthermore, the techniques used in this paper could be used to develop similar algorithms for the CRAY and other machines.

Acknowledgements

I would like to thank my thesis advisor, B. F. Caviness, for his guidance and many helpful suggestions. I would also like to thank R. W. Johnson for many enlightening discussions concerning the material in this paper. Finally, I wish to thank CRAY Research Inc. for the existence of the X-MP.

References

[1] *CRAY X-MP Series Mainframe Reference Manual.* Cray Research Inc., November 1982. HR-0032.

[2] J. Anderson, W. Galaway, R. Kessler, H. Melenk, and W. Neun. *The Implementation and optimization of Portable Standard Lisp for the Cray.* Technical Report Preprint SC-87-1, Konrad-Zuse-Zentrum fuer Informationstechnik Berlin, January 1987.

[3] George E. Collins and Rudiger Loos. *The SAC-2 Computer Algebra System.* Department of Computer and Information Sciences, Ohio State University.

[4] Jeremy R. Johnson. *Some Issues in Designing Algebraic Algorithms for the CRAY X-MP.* Technical Report 88-02, Univ. of Delaware Center for Mathematical Computation, January 1988. Master's thesis.

[5] M. Lauer. Computing by homomorphic images. In B. Buchberger, G. E. Collins, and R. Loos, editors, *Computer Algebra-Symbolic and Algebraic Computation*, pages 139–168, Springer-Verlag, 1982.

EXACT PARALLEL SOLUTION OF LINEAR SYSTEMS

Gilles VILLARD[1]

Laboratoire TIM3-IMAG
Institut National Polytechnique de Grenoble
46, avenue Félix Viallet
38031 Grenoble Cedex France

Abstract : A lot of algorithms provide the exact integer or rational solution of a linear system. Solution may evidently be obtained by a direct computation; an alternative to reduce the *intermediary coefficients swell* is to use reductions modulo primes and p–adic expansions. We present here the parallel implementations which permit us to compare the two methods when applied to large systems.

INTRODUCTION

Let A and b be a square matrix and a vector of sizes n with integer entries (the rational case can be deduced), two main approaches may be used to compute the rational solution vector x such that Ax=b : direct computation and congruence techniques. This second approach may be realized using reductions modulo a lot of prime numbers combined with the Chinese Remainder Theorem, or using only one modulus p and p–adic expansions. Two similar algorithms using p–adic expansions have been given in [5] and [7]. The corresponding arithmetic cost $O(n^3 \log^2 n)$ (number of operations between fixed sized numbers) given by Dixon in [5] represents a major reduction of the *intermediary coefficients swell* phenomenon which usually occurs and which gives at least $O(n^4)$ for the

[1] This work is supported by the *PRC Mathématiques et Informatique* and by the *Greco Calcul Formel* of the french *Centre National de la Recherche Scientifique* (CNRS).

COMPUTER ALGEBRA AND PARALLELISM ISBN 0-12-221100-6

arithmetic cost [1]. After brief presentations of the parallelizations of the direct resolution and of the Dixon's algorithm which have been detailed in [10] and [13], we present experimental results.

The target architecture is assumed to be a network of processors and of connections between neighbours following a hypercube topology (this could be, in fact, easily extended to any other topology : ring, complete network...). Each processor holds a private memory and communicates by message passing. The global behaviour is of MIMD type [8].

1. THE ALGORITHMS

DIRECT COMPUTATION OF THE SOLUTION

Methods to control the integer coefficient growth during Gaussian elimination have been devised. In these methods divisions are allowed when exact.

Parallel Gaussian elimination may be implemented using, like in the sequential case, a unique pivot row at each step. In fact, the communication cost can be reduced if the eliminations are performed using local pivot rows [3, 4, 13] to avoid the costly broadcasting of a global one. We have studied in [10] the behaviour of the fraction–free algorithms given in [1], and shown that the same simplifications can be obtained in the parallel case provided that the eliminations are performed using well chosen elimination strategies.

The parallel implementation which results are presented below has been carried out by developing an *one–step fraction–free* version of the *pipeline ring algorithm* (described in [4, 11]). Even if its communication cost is not the lowest, it permits an easier control of the coefficient growth [10].

COMPUTATION USING P-ADIC EXPANSIONS

The algorithm given in [5] first consists in computing the inverse of the initial matrix modulo a prime number p, then performing Newton–Schultz like iterations we obtain the p–adic approximations of the components of the solution. A *lifting process* finally permits to get the solution. A parallelization has been given in [13]. The main work of one iteration consists in computing two matrix–vector products, they have been implemented as *all to all* communications in the network (see [9, 12]). The last step, to recover the rational values, do not need any communication.

We may notice that practically, the communication cost of this method may be neglected in comparison with the infinite precicion arithmetic costs. Indeed, only reducted modulo values need to be transmited (see [13]). This avoids the costs induced (see [10]) when communicating variable–length and large integers is necessary (like during a direct resolution). In fact, even the communication cost of direct resolutions will not have, in practice, important effects upon the total cost.

We have tested our parallel implementations to calculate the solutions of problems involving matrices with k-digit random elements, with values of k from 2 to 6. So within a factor at most 4/3 (in fact the one–step cost over the two-step cost [1]) we are for the direct resolution in a context analogous to the one used for the timing results given in ([2], table III).

2. EXPERIMENTAL RESULTS

The algorithms have been implemented on the Floating Point Systems parallel computer [6]. It consists of 16 processors connected in a hypercube topology. Each processor is composed of a control and communication unit (Transputer, 7.5 Mips), of a vector processing unit and of a fast memory (1 Mbytes). Codes are written in standard C extended with some FPS communications routines. An *infinite precision arithmetic* has been written on each transputer (see [10] for further details). To *manage memory* on each node we use the C primitives *malloc* and *free*.

EFFICIENCY

The quantity usually considered to compare different parallel algorihtms is the *efficiency* : the ratio of the sequential execution time (i.e. on one processor) to the product of the parallel execution time by the number of processors. This definition not expresses that the total number of performed operations does not depend on the number of processors.

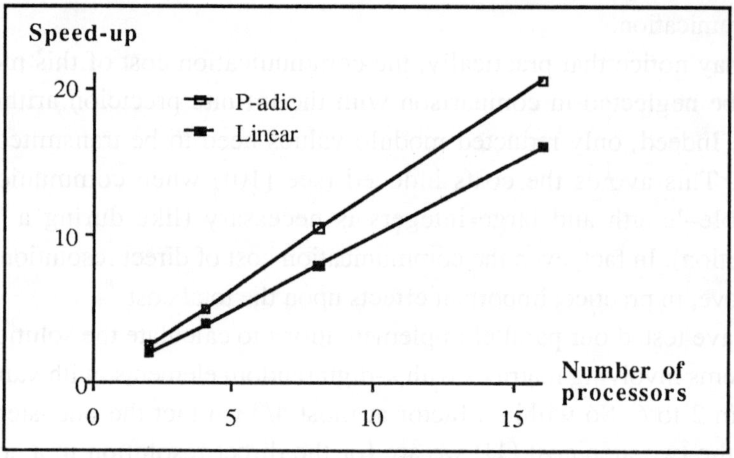

Figure 1 : Speed-up measurements for a 128*128 matrix (entries bounded by 3000).

This ratio is normally expected to be less than one. It would be so if the real cost of an algorithm was solely due to arithmetics and communications. This hypothesis may not be verified in practice : an important extra–cost arises in algebraic computations, the variable–length arithmetic implies a memory management cost. This cost evidently depends on the size of the available memory : it is more saturated during the sequential execution (the total managing cost will be more important) than during the parallel one. This lead to *speed–up* greater than the number of used processors (see figure 1).

DIRECT AND P-ADIC RESOLUTIONS : MEMORY

We call *total managing work* of an execution, the memory that would be necessary to store all the intermediary coefficients. It may be very large : 3,7 GigaBytes for a 256*256 matrix which entries are bounded by 100 and for a direct resolution. For the same matrix and a p–adic resolution we get 0,3 GigaBytes (1,5 GB for a 320*320 matrix).We give on figure 2 the ratios of the total managing works for the two methods. It can be notice that ratios grow quite linearly.

Figure 3 shows us the maximum *memory needed* for each execution. Essentialy the storage of the triangular matrix in the direct case. The storage of two matrices (A and A^{-1} mod p) and of two intermediary vectors in the p–adic case. For small matrices the requirements of the direct method are lower but they grow much faster : when the matrix is of size 320, we need 5,9 MegaBytes versus 4 MB.

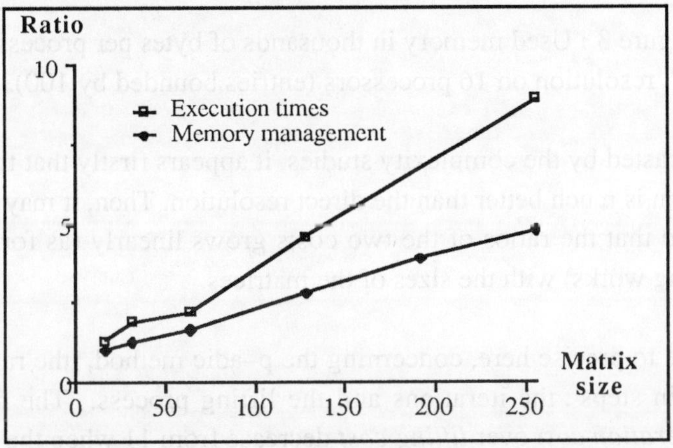

Figure 2 : direct resolution over p-adic resolution, ratios for the total managing work and for the execution times, (entries bounded by 100, 16 processors).

DIRECT AND P-ADIC RESOLUTIONS : EXECUTION TIMES

Assuming we use a classical multiple–precision arithmetic, we recall here the theoretic sequential arithmetic costs : $O(n^5B^2)$ for Bareiss' algorithm [1], where B is a bound on the number of digits of the entries of (A, b), and $O(n^3\log^2 n + n^3 B\log n)$ for Dixon's algorithm [5]. Ratios of the execution times for the two methods are shown on figure 2.

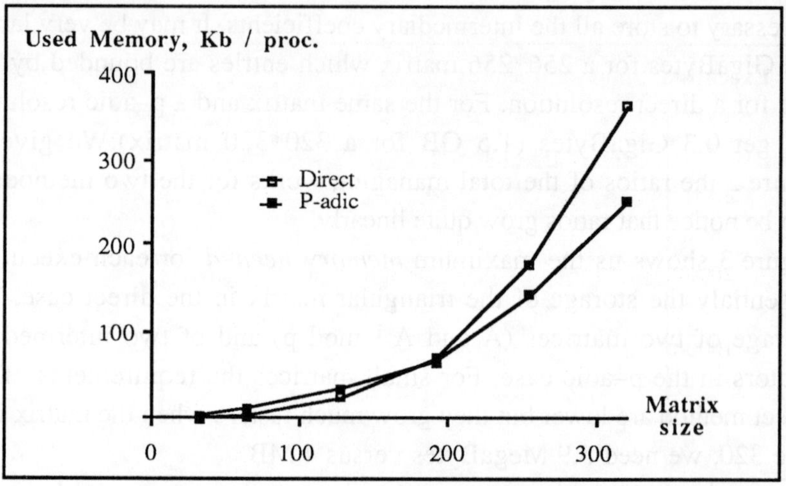

Figure 3 : Used memory in thousands of bytes per processor, resolution on 16 processors (entries bounded by 100).

As forecasted by the complexity studies, it appears firstly that the p–adic resolution is much better than the direct resolution. Then, it may be easily observed that the ratios of the two costs grows linearly (as for the total managing works) with the sizes of the matrices.

We have to precise here, concerning the p–adic method, the ratio of the two main steps : the iterations and the lifting process. The measured ratios *iteration cost* over *lifting cost* decrease from 11 when the matrix is of size 64 to 5 when it is of size 320. The lifting cost does not yet fit the theory. If the complexity to recover one rational is theoretically $O(n^2 \log^2 n)$, the measurements grow like $O(n^3)$: the loss of time essentially come from an important part of the memory management which is done at the end of the algorithm and certainly from a large constant in the big O. And so we have not the $O(n^2 / \log^2 n)$ expected complexity on the figure 2.

Contrary to that, it is interesting to see on the figure 4 the execution times versus the number of digits of the entries. The curves well correspond to the complexities O(B^2) and O(B).

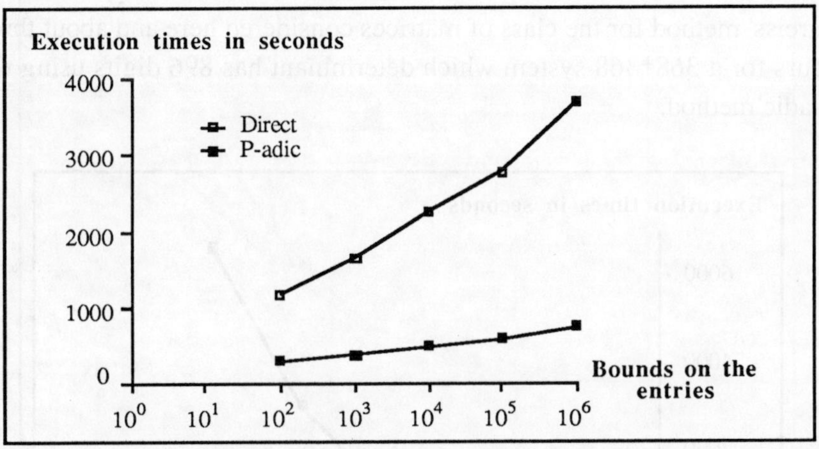

Figure 4 : Execution times versus the sizes of the entries, for a 128*128 matrix (16 processors).

The two last figures give us the execution times, they clearly show the difference between the two methods.

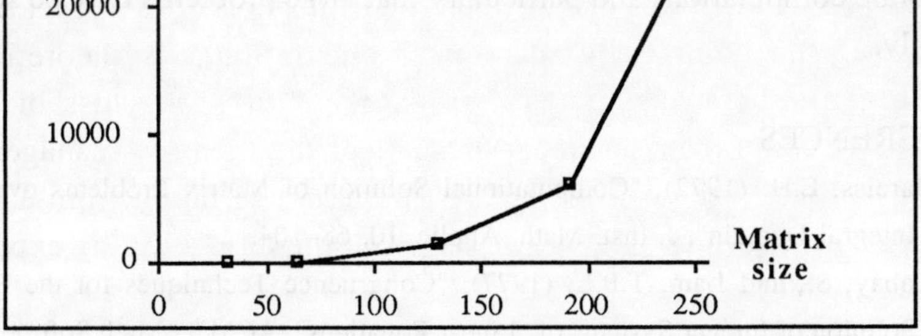

Figure 5 : Execution times of the direct resolution, (entries bounded by100).

CONCLUDING REMARKS

The problem sizes have been increased until the memory was saturated (n=400). It takes about eight hours to solve a 256*256 system using Bareiss' method for the class of matrices considered here and about three hours for a 368*368 system which determinant has 896 digits using the p–adic method.

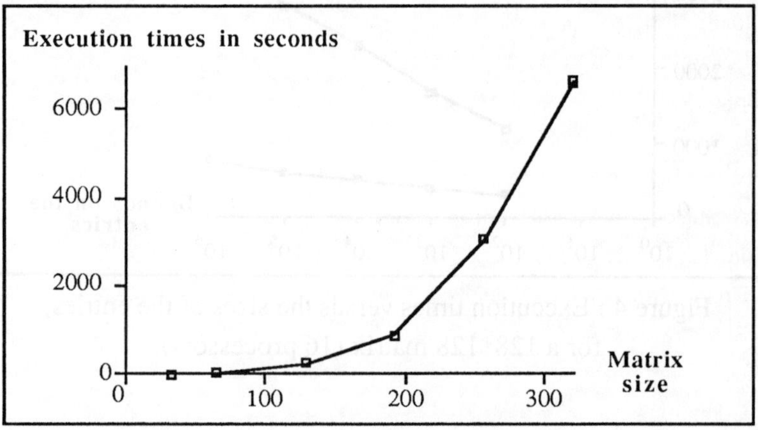

Figure 6 : Execution times of the p–adic resolution, (entries bounded by 100).

This may explain, as said before, some differences between theory and experiments. Such tests would have been quite impossible on a conventional sequential system and would have needed days of computations. We hope so to have shown the parallelism interest in algebraic computations and particulary that large problems may be solved exactly.

REFERENCES

1. Bareiss, E.H. (1972), "Computational Solution of Matrix Problems over an Integral Domain", J. Inst. Math. Applic. 10, 68–104.

2. Cabay, S., and Lam, T.P.L. (1977), "Congruence Techniques fot the Exact Solution of Integer Systems of Linear Equations", ACM Tr.Math.Soft. 3.

3. Cosnard, M. and Robert, Y. (1987), "Implementing the Null Space Algorithm

over GF(p) on a Ring of Processors", Second international symposium on Computer and Information Sciences, Istanbul .

4. Cosnard, M., Tourancheau, B., and Villard, G. (1988) "Gaussian Elimination on Message Passing Architectures", Proceedings of ICS 87, Athènes, Lect. Notes Comp. Sc. no 297, Springer Verlag.
5. Dixon, J.D. (1982) "Exact solution of Linear Equations using P–adic Expansions", Numer. Math. 40, 137–141.
6. Floating Point Systems, "Programming the FPS T–Series, Release B, Portland Oregon 97223.
7. Gregory, R.T., and Krishnamurthy, E.V. (1984) "Methods and Applications of Error–Free Computations", Springer Verlag.
8. Hwang, K., and Briggs, F. (1984) "Parallel Processing and Computer Architecture", Mc Graw Hill.
9. Johnsson, S.L., and Ho, C.T. (1986) "Spanning Graphs for Optimum Broadcasting and Personalized Communication in Hypercubes", Technical Report 500, Comp. Sc. Dpt.,Yale University.
10. Roch, J.L., Sénéchaud, P., Siebert, F., et Villard, G. (1987) "Parallel Algebraic Computing", Imag Grenoble, RR–686 I.
11. Saad, Y. (1986) "Gaussian Elimination on Hypercubes", in Parallel Algorithms and Architectures, Eds. M.Cosnard & al., North-Holland.
12. Stout, Q.F., and Wager, B. (1987) "Intensive Hypercube Communication I : Prearranged Communication in Link–Bound Machines", CRL–TR–9–87, University of Michigan.
13. Villard, G. (1988) "Parallel General Solution of Rational Linear Systems using P–adic Expansions", Proceedings of the IFIP WG 10.3 Working Conference on Parallel Processing, Pisa Italy, Elsevier Sc.P.

SOLVING LINEAR DIOPHANTINE EQUATION IN PARALLEL

Françoise SIEBERT-ROCH[1]

Laboratoire TIM3 - IMAG
Institut National Polytechnique de Grenoble
46, avenue Félix Viallet
38031 Grenoble Cedex France

Abstract : Find out the Hermite normal form of a matrix A is a way to solve the system Ax = b for integral value of x, where A is an integral matrix and b an integral column vector.

Solving integral linear systems in Computer Algebra requires to handle a large number of integers and the algorithms show some important intermediate coefficient swell. These factors have lead us to consider a parallel implementation.

This paper is dedicated to a particular problem : solving a linear Diophantine equation.

In the first part we give a brief theoretical overview of that subject.Then we present an algorithm described by W.A.Blankinship [BLA67] determining the GCD d of n positive integers $a_1, a_2, \ldots, a_n$ and giving a solution $x_1, x_2, \ldots, x_n$ for the Diophantine equation:

$$d = a_1 x_1 + a_2 x_2 + \ldots + a_n x_n.$$

This algorithm has been parallelized and implemented on a parallel machine of MIMD type. Some experimental results are presented in the last part of the paper.This work is a first step in the parallelization study of algorithms as Hermite, Lenstra, … .

---

[1] This work is supported by the *PRC mathématiques et informatique* and by the *GRECO calcul formel* of the french *Centre National de la Recherche Scientifique* (CNRS).

COMPUTER ALGEBRA AND PARALLELISM ISBN 0-12-221100-6

INTRODUCTION

We study the general problem of integral resolution of linear systems. That is :

To find $x \in \mathbb{Z}^n$ such that : $\mathbf{A}\,\mathbf{x} = \mathbf{b}$

where A is an m x n matrix of full row rank with coefficients in $\mathbb{Z}$ and b a vector from $\mathbb{Z}^m$.

The problem can be solved by the Hermite method:

Theorem : for all the matrix A (m, n) of rank m ($m \leq n$) with elements in $\mathbb{Z}$, there exists an unimodular matrix U such that :

$$AU = H$$

where H is regular and triangular inferior and his entries satisfie:

$$0 \leq h_{ij} < h_{ii} \; (j < i \leq m).$$

H is the Hermite normal form of A.

From H, the solution of the system (if it exists) can be calculated directly.

The system $\mathbf{A}\,\mathbf{x} = \mathbf{b}$ is in fact a system of Diophantine equations of the form

$$a_{i1}\, x_1 + a_{i2}\, x_2 + \ldots + a_{in}\, x_n = b_i.$$

Here we shall be concerned with the particular case of one Diophantine equation. Let us first consider the problem theoretically, and then study one particular algorithm. The algorithm proposed by W.A. Blankinship [BLA67], calculates the a_i's GCD d and the general solution of :

$$a_1\, x_1 + a_2\, x_2 + \ldots + a_n\, x_n = m,$$

for all multiple m of d.

The parallel architectures allow to treat some problems with high granularity. So, using such computers looks suitable for large general

computer algebra problems composed of multiple independent tasks. Lots of linear algebra problems meet this criterion. In particular, the parallel approach may be applied for our specific problem.
We propose a parallel algorithm solving a Diophantine equation, adapted to a particular parallel model and more precisely to the FPS T20 machine.
In the last part we shall present experimental results obtained on this machine.

I - THEORETICAL APPROACH

Let the equation :

$$(1) \qquad a_1 x_1 + a_2 x_2 + \dots + a_n x_n = b$$

with $a_1, a_2, \dots a_n$, b integers. This equation has an integral solution if and only if the a_i's GCD divides b.
Consider the application :

$$\phi : \quad \mathbb{Z}^n \longrightarrow \mathbb{Z}$$
$$(x_1, x_2, \dots, x_n) \mapsto a_1 x_1 + a_2 x_2 + \dots + a_n x_n$$

ϕ is an homomorphism of $\mathbb{Z}$-module. The kernel of ϕ is a sub-module of $\mathbb{Z}^n$ of dimension n - 1. It exists n - 1 independent solutions of the homogeneous equation $s_1, s_2, \dots, s_{n-1} \in \mathbb{Z}^n$ such that:

$$\mathrm{Ker}\,\phi = \{\lambda_1 s_1 + \lambda_2 s_2 + \dots + \lambda_{n-1} s_{n-1} / \lambda_1, \lambda_2, \dots, \lambda_{n-1} \in \mathbb{Z}\}.$$

$s_1, s_2, \dots, s_{n-1}$ form a basis of the $\mathbb{Z}$-module Ker ϕ.

Theorem :

Let b be the GCD of the a_i (i = 1, ...n) , and $x_0 \in \mathbb{Z}^n$ a particular solution of (1). The general solution $x \in \mathbb{Z}^n$ of (1) can be put on the form :

$$x = x_0 + \lambda_1 s_1 + \lambda_2 s_2 + \dots + \lambda_{n-1} s_{n-1} \quad \text{with} \quad \lambda_1, \lambda_2, \dots, \lambda_{n-1} \in \mathbb{Z}.$$

Demonstration :

If x is a solution of (1) we have $\phi(x - x_0) = 0$. So $x - x_0 \in \mathrm{Ker}\,\phi$.

If b = c d with c integer then the general solution of (1) is (see [SCH85]):

$$(2) \quad x = (x_0 * c) + \lambda_1 s_1 + \lambda_2 s_2 + \ldots + \lambda_{n-1} s_{n-1}$$

If we have the general solution for $|a_1|, |a_2|, \ldots, |a_n|$, the general solution for $a_1, a_2, \ldots, a_n$ is obtained by multiplying each column of the former general solution by the sign of the corresponding a_i. Indeed:

if $$|a_1|\, x_1 + |a_2|\, x_2 + \ldots + |a_n|\, x_n = m,$$

then $a_1(\varepsilon_1 x_1) + a_2(\varepsilon_2 x_2) + \ldots + a_n(\varepsilon_n x_n) = m$ where $\varepsilon_i = \mathrm{sign}(a_i)$.

Then we shall assume the a_i's positive.

II - ALGORITHMIC APPROACH

II-1 Algorithm description

Let $D = (d_{ij})_{(i = 1 \text{ to } n,\ j = 1 \text{ to } n+1)}$ be the n x (n + 1) matrix

$$\begin{bmatrix} a_1 & 1 & 0 & \ldots & 0 \\ a_2 & 0 & 1 & \ldots & 0 \\ & & \ldots & & \\ a_n & 0 & \ldots & 0 & 1 \end{bmatrix}$$

. Let A be the first column vector of D and B the matrix constructed with the last n columns of D.

We perform elementary transformations on the rows of D so that their first coefficients decrease and their GCD is not modified. These transformations correspond to Euclidean divisions on the d_{i1} ($i = 1, \ldots, n$). d_{i1} is always divided by a smaller d_{k1}. This process will be iterated until all the first column coefficients are zeros except one : let $d_{i_0 1}$ be this coefficient, and **D** the matrix obtained.

The d_{k1} decrease and are minored by 0, this accounts for the convergence of the algorithm. At the beginning we have :

$$\mathbf{A} = \mathbf{B} \times \mathbf{A} \quad (\mathbf{B} = \mathbf{I}_n)$$

If $\bar{\mathbf{B}} = (\bar{b}_{ij})_{i=1,\dots,n}$ is the matrix built with the last n columns of **D**, we obtain :

$$\begin{bmatrix} 0 \\ 0 \\ . \\ d_{i_0 1} \\ 0 \\ . \\ 0 \end{bmatrix} = \bar{\mathbf{B}} \times \mathbf{A}$$

, where $d_{i_0 1}$ is the a_i's GCD (See [MAL80]).

The i_0^{th} row of $\bar{\mathbf{B}}$: $(\bar{b}_{i_0 1}, \bar{b}_{i_0 2}, \dots, \bar{b}_{i_0 n})$, is a particular solution for the Diophantine equation. The other rows give (n - 1) solutions for the homogeneous equation. The general solution depends on (n - 1) parameters $\lambda_1, \lambda_2, \dots, \lambda_{n-1}$. The mth variable (m = 1, ..., n) is given by : (1) $x_m = \bar{b}_{i_0 m} + \lambda_m \sum_{i \neq i_0} \bar{b}_{i m}$, $\lambda_m \in \mathbb{Z}$.

The process to cancel first column coefficients can be performed in different ways. We describe here two sequential algorithms.

S1 Algorithm

Data inputs: n positive integers $a_1, a_2, \dots, a_n$

Step 1: $D \leftarrow\!\!-\!\!- \begin{bmatrix} a_1 \; 1 \; 0 \dots 0 \\ a_2 \; 0 \; 1 \dots 0 \\ \dots \\ a_n \; 0 \; \dots 0 \; 1 \end{bmatrix}$

Step 2: $d_{k1} \leftarrow\!\!-\!\!- \mathrm{Min}_{i=1,\dots,n} (d_{i1} , d_{i1} \neq 0)$;

if D has only one non-zero leading coefficient, go to step 5.

Step 3: For $i \neq k$ do

if $d_{i1} \neq 0$ then

q <--- d_{i1}/d_{k1} ;

d_{ij} <--- $d_{ij} - q\, d_{kj}$ $\quad (j = 2, \ldots, n+1)$;

endif

endfor

Step 4: Go to Step 2.

Step 6: d_{k1} is the gcd of the given integers and the kth row of D gives asolution for the Diophantine equation.

This algorithm consists in choosing for each division one of the smallest non zero remainders. The same transformation is performed on all the rows (excepted the one corresponding to the smallest non zero remainder chosen and these with zero first column coefficient). The leading coefficients decrease rapidly, but at each step the search for the smallest leading coefficient is to be performed.

S2 Algorithm

Step 1: Idem

Step 2: d_{k1} <--- $\mathrm{Min}_{i=1,\ldots,n}(d_{i1}\,,\, d_{i1} \neq 0)$

if D has only one non-zero leading coefficient, go to step 4.

Step 3: While it exists d_{i1} so that $i \neq k$ and $d_{i1} \neq 0$ do

3.a : (c_0, c_1) <-- (d_{i1}, d_{k1}) ;

(s_0, s_1) <-- $(1, 0)$; (t_0, t_1) <-- $(0,1)$;

3.b : q <--- c_0/c_1 ;

(c_0, c_1) <-- $(c_1, c_0 - q\, c_1)$;

(s_0, s_1) <-- $(s_1, s_0 - q\, s_1)$; (t_0, t_1) <-- $(t_1, t_0 - q\, t_1)$;

3.c : if $c_1 \neq 0$ then

(k, i) <-- (i, k) ;

<u>go to</u> step 3.b ;
<u>else</u> (d_{ij}, d_{kj}) <-- ($(d_{k1}\, d_{ij}) /c_0$ - $(d_{i1}\, d_{kj}) /c_0$,
$s_0\, d_{ij} + t_0\, d_{kj}$) (j = 2, …, n+1) ;
(d_{i1}, d_{k1}) <-- $(0, c_0)$;
<u>endif</u>

<u>endwhile</u>

Step 4: d_{k1} is the gcd of the given integers and the kth row of D gives a solution for the Diophantine equation.

A non zero first column coefficient is selected, the transformations between this coefficient and an other one non zero first column coefficient is perfomed until the remainder is zero (we perform the Euclidean scheme between the two coefficients). We perform the corresponding transformation on the rest of the two rows concerned. The non zero first column coefficient obtained is now the smallest one. Less comparisons are performed than in the former algorithm - no minimum search on the first column at each step.
On each row the last columns initialized to identity matrix contain, at step k, at least n - k zeros. This algorithm allows to cover the matrix by increasing order of row indices (except the row indice corresponding to the first coefficient selected). So, we know the positions of the n - k zeros, corresponding to the n - k unmodified rows. We can take these zeros into account at the programming level.

Former algorithms, both allow us to recover the general solution of the Diophantine equation using the relation (2).

II-2 Parallel Model

The chosen parallel model is an architecture of MIMD type (there is no global synchronism between processors) with high granularity. Each processor has a local memory. It allows us to perform huge

computations and to store very large matrices. For this algorithm the absence of shared memory is not an handicap, each processor works at each step on independent data.
The processors communicate with their neighbours by physical links. The transfer time (sending or receiving) of n consecutive data to a neighbouring processor is defined by [SAA85] :

$$\beta + n\,\tau$$

Where β is the initialization time, and τ the transfer time.

II-3 Parallel Algorithm

We first remark that the gcd computation presents an inherent parallelism: the gcd of $(a_1, a_2, \ldots, a_n, \ldots, a_{2n})$ is also the gcd of (d_1, d_2) where d_1 is the gcd of $(a_1, \ldots, a_n)$ and d_2 the gcd of $(a_{n+1}, \ldots, a_{2n})$. The determination of d_1 and d_2 can be performed in parallel on two processors. The problem can be broadcast homogeneously in several processors: performing the algorithm described before on submatrices, we obtain in each processor only one row with non zero first column coefficient; then we continue the computation on these rows.
The parallelization will consist in choosing a good sequential method on each processor, and performing the process on submatrices.
The number of coefficients of the considered matrices grows quadratically with the number of terms whose gcd is to be determined. So, the size of the manipulated objects increases fastly and can lead to a memory overflow.
The parallel programming allows a manipulation of smaller matrices and so the data size decreases quadratically with the number of processors.
Working on p processors we have :

$$\text{data number on each processor} = (n/p) * (n/p + 1)$$

and so :
(p*Data number on each processor) << Data number for the sequential problem!!!

Once the algorithm has been performed on each submatrices on their assigned processor, the size of the obtained rows must be doubled by completing with zeros and moving, if necessary, all the coefficients but the first one.

The submatrices of the global matrix are evenly distributed on the processors. In two processors communicating at the first step, we must have two adjacent matrices in order to perform the transformations on the two rows obtained. In the same way, in the following steps, when a processor transmits a row to a neighbour *proc*, it has to be issued from the consecutive submatrix (in D) of the submatrix considered in the processor *proc*.

In the following, we shall assume that the labels of two neighbouring processors differ from one single bit in their radix 2 representations.

Once the first step of transformation on the submatrices performed, the row transfer process chosen is the following one :

At step k (k : 1, …, dim) : the processor whose radix 2 representation is $(1, b_{dim\text{-}k\text{-}1}, \ldots, b_1, b_0)_2$ sends the obtained row to the processor $(0,b_{dim\text{-}k\text{-}1},\ldots,b_1,b_0)_2$ (figure1).(See[SAA85]).

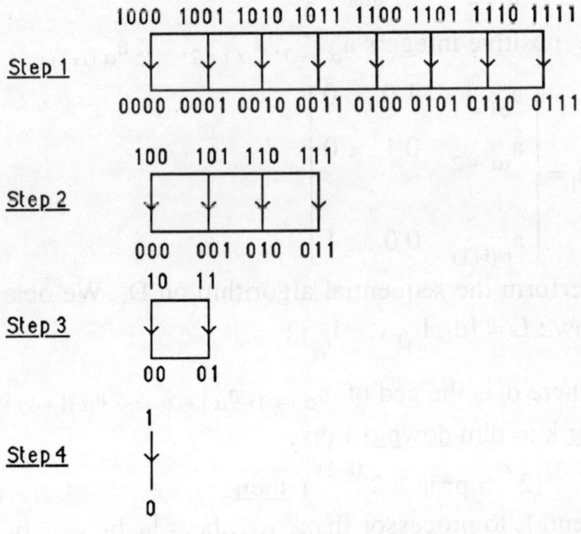

Fig 1 : Lifting process to processor 0 for 16 processors

If the number of coefficients is $n = 2^e$ and the number of used processors is $p = 2^{dim}$ (dim < e) then, to obtain at the end of the algorithm (in processor 0) the row containing the GCD and the resulting coefficients set in the initial order, we have to distribute the submatrices in the following way :

The processor *proc* = $(b_{dim-1}, b_{k-1}, \ldots, b_1, b_0)_2$ receives the coefficients corresponding to the row and column indices :

$$\beta = \left(\sum_{k=1}^{dim} \frac{b_{k-1}}{2^k}\right) n + 1 = \left(\sum_{k=1}^{dim} b_{k-1} 2^{dim-k}\right)\left(\frac{n}{p}\right) + 1 \text{ -----> } \beta + \frac{n}{2^{dim}}$$

P Algorithm (on $p = 2^{dim}$ processors)
If processor=*proc* = $(b_{dim-1}, \ldots, b_1, b_0)_2$:

Let: $\alpha = n / p$;
$i = (b_0, b_1, \ldots, b_{dim-1})_2$;

Inputs: α positive integers $a_{\alpha i+1}, a_{\alpha i+2}, \ldots, a_{\alpha(i+1)}$;

Step 1:

$$D_i = \begin{bmatrix} a_{\alpha i+1} & 1\,0 \ldots 0 \\ a_{\alpha i+2} & 0\,1 \ldots 0 \\ \ldots & \ldots \\ a_{\alpha(i+1)} & 0\,0 \ldots 1 \end{bmatrix}$$

Step 2: Perform the sequential algorithm on D_i. We obtain one row : $L = [d_i, l_{i_1}, \ldots, l_{i_\alpha}]$;

where d_i is the gcd of $a_{\alpha i+1}, a_{\alpha i+2}, \ldots, a_{\alpha(i+1)}$;

Step 3: for k := dim downto 1 do :

If $(2^k > \text{proc} \geq 2^{(k-1)})$ then

Send L to processor $(b_{dim-1}, \ldots, b_{k+1}, \bar{b}_k, b_{k-1}, \ldots, b_0)_2$;

else if $(2^{(k-1)} > \text{proc} \geq 0)$

Let R a $(\frac{n}{2^{k-1}} + 1)$ dimension vector;

Initialize $R := \left[d_{i'}, 0, \ldots, 0, l_{i'_1}, \ldots, l_{i'_\alpha} \right]$ where $\left[d_{i'}, l_{i'_1}, \ldots, l_{i'_\alpha} \right]$ is the row send by processor $(b_{dim-1}, \ldots, b_{k+1}, \bar{b}_k, b_{k-1}, \ldots, b_0)_2$;

Let $S := \left[d_i, l_{i_1}, \ldots, l_{i_\alpha}, 0, \ldots, 0 \right]$ (S equal L completed with α zeros) ;

Perform elementary transformation between S and R ;

If $R[1] = 0$ then $L := S$ else $L := R$;

endif

$\alpha := 2 * \alpha$;

Step 4: The result is recovered in processor 0.

Let us notice that, provided that the transformed rows in step 3 are preserved, we can construct the general solution. We can recover the total transformed matrix at the end of the algorithm - coefficients non stored are all zeros -.

II-4 Complexity

Let us assume that we work on 2^{dim} processors. Let $\alpha = n/ 2^{dim}$, and we note $|A| = \max_{i=1,\ldots,n} (|a_i|)$.

In step 2, in processor p, we compute in sequential the GCD of α integers. This computation can be performed in $O(\alpha \lg |A|)$ operations (the basic operation is supposed to be division) . The associated row operations require $O(\alpha^2)$.

The cost of step 3 is the cost of :

- the GCD computation of dim integers, that is $O(\dim \lg |A|)$
- $O(\alpha 2^K)$ row operations and $((2^{k-1}\alpha + 1) N_k \tau + \beta)$ communication time at step k .

We note N_k a bound on the coefficients size : $N_k \sim O(|A|^{\alpha k})$
The total communication time is

$$\sum_{k=1}^{\dim} (2^{k-1}\alpha + 1) N_k \tau + \dim \beta$$

and the computation complexity is of order : $O(\alpha^2 + \alpha\, 2^{\dim})$

III - EXPERIMENTAL RESULTS

The algorithm has been implemented on the TESSERACT 20 of Floating Point System, MIMD computer with hypercube design .

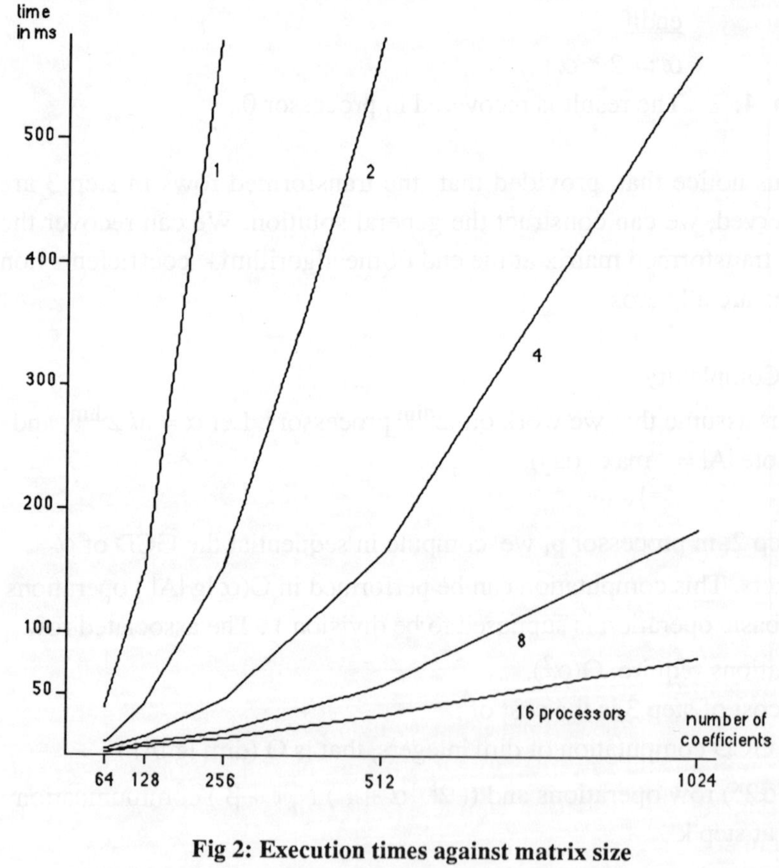

Fig 2: Execution times against matrix size

The algorithms have been programmed in the language OCCAM with primitives allowing for the data transfer between processors.

β varies between 9.5 µs and 60 µs (it depends on the number of used links) and τ is about 6 µs for an integer coded on 32 bits.

For more precision on the machine see [CTV87].

The performances obtained (Fig 2) clearly show the computation time saving allowed by the parallelism. In most cases, we even obtain time reductions by a factor higher than two, when the number of processors is only doubled. The algorithm programmed on two processors is not the same than on one processor, the matrices treated at the first step are smaller as well as the total number or performed operations.

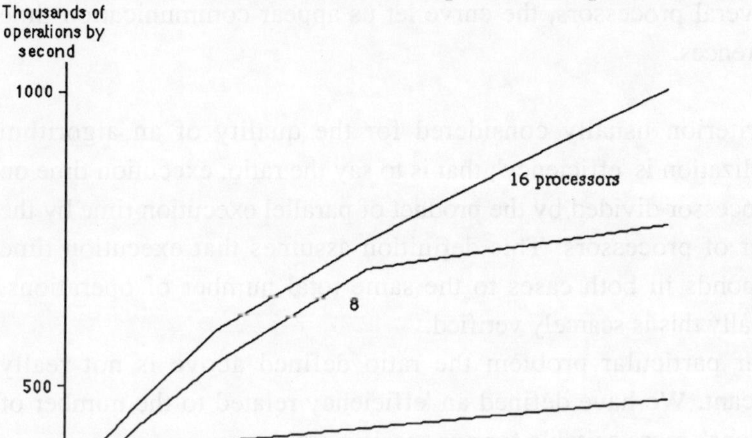

Fig 3 : Number of operations per second

We communicate rows, vectors can be send by block and so we have only one initialization time. Moreover, the transfered objects (rows) are small in comparisons with the manipulated objects (matrices).

Let us notice that, for a large number of coefficients, it is necessary to use several processors to perform the computation. Each processor of the FPS T20 has a memory of one Mbyte. When n grows up, the avaiable memory on one node is fastly unsufficient to store the total matrix : so, on one node, the tests have been only performed for problems of size lower than 256.

The figure 3 represents the performances obtained, in thousands of operations per second.
For one processor we have a straight line, there is no communication and so the computation power is the same whatever the data number.
For several processors, the curve let us appear communication time interferences.

The criterion usually considered for the quality of an algorithm parallelization is 'efficiency'; that is to say the ratio: execution time on one processor divided by the product of parallel execution time by the number of processors. This definition assumes that execution time corresponds in both cases to the same total number of operations. Practically this is scarcely verified.
For our particular problem the ratio defined above is not really significant. We have defined an 'efficiency related to the number of operations', more suitable for our specific problem.

Let e be the ratio : $e = \dfrac{\dfrac{\text{number of operations on p processors}}{\text{// time * p}}}{\dfrac{\text{number of operations in sequential}}{\text{sequential time}}}$

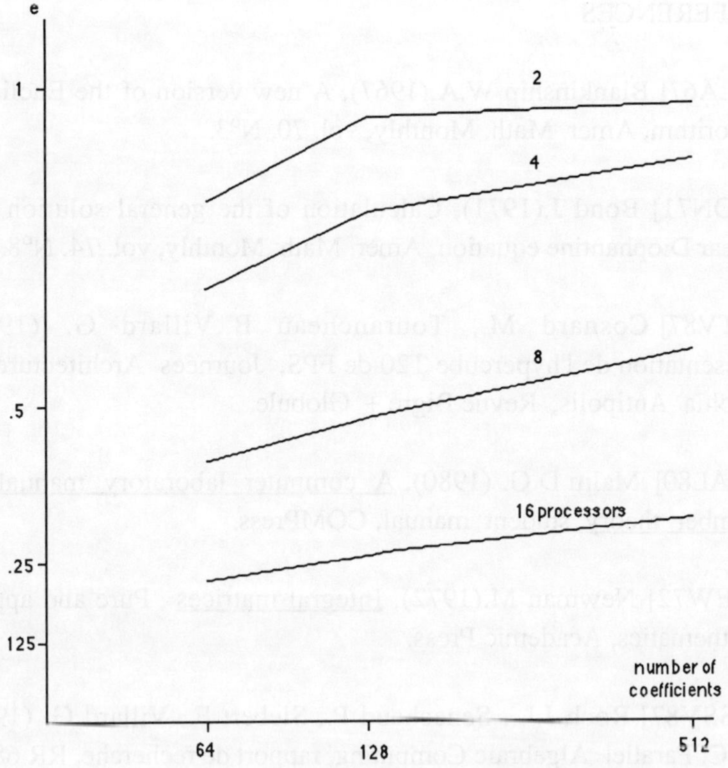

Fig 4: Efficiency related to the number of operations

CONCLUSION

The parallelization of the algorithm has brought both time and storage improvements.

Let us remark that the sequential algorithm is thought sequential. The simulation of the parallel algorithm on one processor, in view of the results obtained, would have improved the sequential algorithm.

The good performances obtained let us hope that some algorithms like Hermite method could be efficiently parallelized.

REFERENCES

[BLA67] Blankinship W.A.(1967), A new version of the Euclidean algorithm, Amer. Math. Monthly, vol. 70, N°3.

[BON71] Bond J.(1971), Calculation of the general solution of a linear Diophantine equation, Amer. Math. Monthly, vol. 74, N°8.

[CTV87] Cosnard M., Tourancheau B.,Villard G. (1988), Présentation de l'hypercube T20 de FPS, Journées Architecture C3, Sophia Antipolis, Revue Bigre + Globule.

[MAL80] Malm D.G. (1980), A computer laboratory manual for number theory, student manual, COMPress.

[NEW72] Newman M.(1972), Integral matrices , Pure and applied mathematics, Academic Press.

[RSSV87] Roch J.L., Senechaud P., Siebert F., Villard G. (1987), PAC: Parallel Algebraic Computing, rapport de recherche, RR 686-I.

[SCH85] Schrijver A.(1985), Theory of linear and integer programming , John Wiley, Chichester, England.

[SAA85] Saad Y. (1985), Topological properties of hypercubes, Research report YALEU / DCS / RR-389.

Parallel Algorithm Development

Steffen Seitz
Wilhelm Schickard Institut für Informatik
Universität Tübingen

Abstract

Using the SAC-2 computer algebra system on a network of SUN Workstations [1], we show how the user can utilize several machines for the execution of a program simply by setting the unit number for input and output. Our approach is demonstrated by a distributed version of the algorithm IPRES[1] from the GCD and Resultant Package of the SAC-2 library.[2]

1 Motivation

The modular representation of integers as sets of r moduli allows the calculation of some arithmetical operations on two integers by r operations on r moduli. The r operations can be executed separately and independently at the same time. The same holds for the modular representation of other objects, e.g. polynomials, where the modular operations are performed for the coefficients.

Non-algebraic algorithms, which do not rely on the algebraic algorithms of SAC-2 but use its symbol and list processing capabilities, can be splitted into independent subalgorithms. The generation of ALDES algorithms from the specification of a term rewriting system[4] can be done independetly for every defined operator.

2 Environment

2.1 System

Research in an environment with limited resources is usually governed by several 'axioms'[2].

Our current situation does not allow the development of new hardware, which might be specialized for applications in computer algebra, thus we have

[1] Sun Workstation is a registered trademark of Sun Microsystems, Inc.
[2] This paper contains material of a forthcoming Ph.D. Thesis supervised by R. Loos

COMPUTER ALGEBRA AND PARALLELISM ISBN 0-12-221100-6

to use stock hardware. The same holds for system software as operating systems and the implementation of robust network protocols.

The hardware environment consists of SUN Workstations connected by an Ethernet, running the SUN version 3.5 of the UNIX[3] operating system. The socket mechanism is used for the transmission of data on the network [5].

The development of a computer algebra system together with a language, which offers concepts for the formulation of distributed algorithms or at least offers the flexibility to add such concepts, is a non-trivial project. Even the usage of some existing systems to write such algorithms to is not easy, because they are either closed, i.e., no source code is available, or the algorithms are not well documented.

Our software environment is formed by the computer algebra system SAC-2 because much knowledge about its internal structure and its usage is available on the local site and it is the only open system with all documented sources available [3].

2.2 Model of Parallelism

Our study of parallelism in algebraic algorithm is concerned with and restricted to high level, problem oriented, coarse grained parallelism. The algorithms are intended to run on a multicomputer, i.e., a model of a MIMD computer with no shared global memory. Each processor has its own local memory, and process communication occurs through message passing. The time for the transmission of data between hosts should be small in contrast to the time spent on each host processing them. In addition the dynamic creation and deletion of processes is not yet considered.

3 Present state

Figure 1 shows how data is read and written in the SAC-2 system. Each algorithm in the SAC-2 system, which wants to get data displayed on the terminal or written into a file, uses algorithms which ultimately put single characters, by the algorithm CWRITE, into an array buffer called OBUFF. When the last position in OBUFF has been used, CWRITE calls WRITE, which translates the character code and executes an output operation for the transfer of the content of OBUFF to a local device, specified by the global variable OUNIT.

Analogously, an input operations fills the input buffer IBUFF with data from the device specified by IUNIT. The input operation is executed by the algorithm READ, as soon as the algorithm CREAD detects IBUFF empty. Procedures, which implement the input and output operation are easily written. They just have to pass, for example by simple system calls, the content of OBUFF to local devices and vice versa.

[3]UNIX is a trademark of AT&T Bell Laboratories

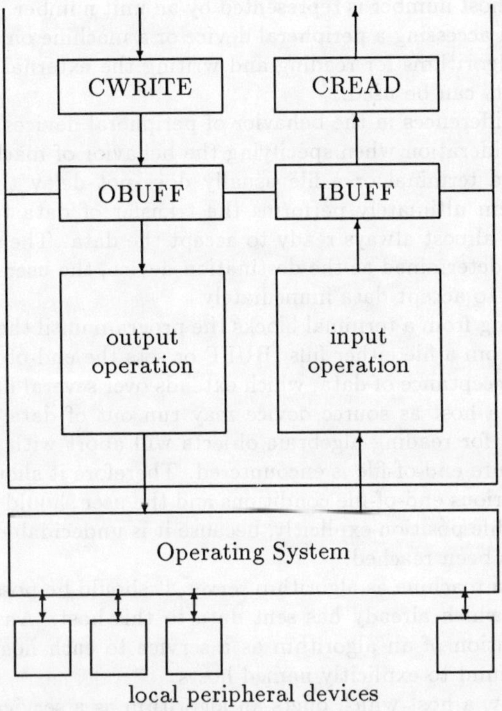

Figure 1: Reading and writing in SAC-2

4 Approach

We exploit the fact, that the SAC-2 system only knows about streams of characters coming from and flowing to devices which are identified by logical unit numbers. This concept allows the denotation of virtually any physical or logical device in the computer system.

In order to send data to other machines, each host on the network gets a host number. Each host number is represented by an unit number. The users sees no difference when accessing a peripheral device or a machine on the network, thus the standard algorithms for reading and writing the external representation of algebraic objects can be used.

There are differences in the behavior of peripheral devices which have to be taken into consideration when specifying the behavior of machines as devices.

Writing to a terminal or a file usually does not delay the program. The operating system ultimately performs the transfer of data to the destination device which is almost always ready to accept the data. The same should hold when a host is determined as the destination device, the user should see a host being prepared to accept data immediately.

While reading from a terminal blocks the program until the user enters some data, reading from a file either fills IBUFF or sets the end-of-file condition.

When the acceptance of data, which extends over several lines, is faster than its generation, a host as source device may run out of data *for a short time.* The algorithms for reading algebraic objects will abort with an error message when a premature end-of-file is encountered. Therefore it should be possible to ignore such spurious end-of-file conditions and the user should be able to set the genuine end-of-file position explicitly, because it is undecidable when the logical end of data has been reached.
In order to use a machine as algorithm server, it should be possible to read data from any host which already has sent data to this host. An algorithm server offers the execution of an algorithm as a service to each host on the network and it is not bound to explicitly named hosts.

Subsequently, a host which offers an algorithm as a service is called *server*, the *requestor* is a host which calls on a service.

The requestor sends the actual parameters for the algorithm to the server, and it gets the results by reading, virtually, from the server. Results are sent back to the requestor into a local buffer. The requestor starts to read the results at any time. Either data is available in the local buffer and the requestor proceeds or the server has not yet begun to return data and the requestor waits, until data arrives or gets an end-of-file condition. The behavior of the requestor is determined by a global variable BLOCK.

5 Implementation

The view of the hosts on a network, which has been specified in the preceding section, has been implemented on a network of around 20 SUN Workstations and fits within 950 lines of C code.

The code, which already implements the input and output operations with local physical devices is partitioned into sections with each section is responsible for a particular class of devices. Only one section, for the class of local physical devices, is used.

A new section with procedures for data transmission on the network is introduced. The value of IUNIT and OUNIT determine the section to be used for the input and output operation.

Every unit number greater than 100 uniquely denotes a host. The host with host number h corresponds to unit 100+h.

The correspondence between host numbers and network addresses in respect to the addressing convention of the message handling facilities of the operating system has to be established by the user in a text file, which is read at initialization time of a program.

The file, subsequently called the hosttable, consists of lines of the following form

<host_number> <host_address>.

Every text in a line after the character '#' is ignored. Figure 2 shows a possible configuration.

We now describe, how the behavior of hosts as peripheral devices is implemented.

In order not to delay an output operation until the receiving program accepts the data, data is always written to a buffer process which exists on each host on the network and which stores the data in its buffer area together with the host number of the sender. When doing an output operation to an other host, a connection is established to the buffer on the destination host. The operating system provides the means to queue simultaneous connection requests of various hosts.

When reading from a host, the execution of an input operation leads to a connection to the local buffer process. A *non-blocking* input operation either fills IBUFF with data found in the buffer or sets the global variable EOF = TRUE, indicating an end-of-file condition. A *blocking* input operation always waits for the arrival of further data from host with number h, when the buffer does not contain data from host h. The desired kind of input operation is specified by the value of the global variable BLOCK. BLOCK = TRUE forces a blocking, BLOCK = FALSE a non-blocking input operation.

Reading with IUNIT set to 100 is treaded as a special case. The data from any host, which has sent data to the local buffer may be retrieved. The number

```
# Hostnumbers for distributed ALDES Algorithms
#
# Host  Internet Address

   99      129.143.72.180  # SECUNDUS
#          129.143.72.181  # c1
   5       129.143.72.182  # c2
#          129.143.72.183  # c3
   6       129.143.72.184  # c4
   7       129.143.72.185  # c5
   8       129.143.72.186  # c6
#          129.143.72.187  # c7
   1       129.143.72.188  # c8
   2       129.143.72.189  # c9
   3       129.143.72.190  # c10
#          129.143.72.191  # c11
#          129.143.72.192  # c12
   4       129.143.72.193  # c13
```

Figure 2: Sample hosttable

of the actually chosen host is left by the INPUT primitive in the global variable HOST and can be used to send back an answer by setting OUNIT = 100 + HOST.

A particular situation arises, when IUNIT is set to 100 and the user asks for a blocking input operation. After having selected a host from which to read data, the input procedure keeps on reading from this host, even when the end of the data has arrived. The user has to inform the INPUT procedure, that an other host number should be chosen for the retrieval of data.

Because the sender of data knows when he has completed its transmission, it sends a logical-end-of-data record. Whenever the input procedure reads such a record, it 'frees' the current host and the next call for input again selects an arbitrary host. Marking by the sender is done by a call of the procedure FIN after the data has been written.

6 An Example

The example in listing 1 shows the subalgorithm MPRES (Modular Polynomial RESultant) of IPRES (Integer Polynomial RESultant), which can be started in several copies on the network.

In addition to the initialisation of the polynomial package with BEGIN3, BEGINC initializes the network section of the input and output operations by reading the hosttable.

MPRES is called from the main program, which performs the initialization of the polynomial package (BEGIN3) and the network section of the input and output operations by reading the hosttable (BEGINC). The main program then repeatedly reads from *any* host (IUNIT = ANY) the arguments r, p, A* and B* for MPRES. The results C* is written back to the sender of the current arguments (OUNIT = SRV(HOST)).

The host running MPRES, encapsulated in the main program, thus offers the service to compute the resultant of two modular polynomials.

```
safe HOST, BLOCK.

const ANY = 100.                                       [ unspecified host ]
const SRV(h) = 100 + h.                     [ host number to unit number ]

                              main.
[Modular polynomial resultant, server version.
The prime number p, the beta digit r and two polynomials A and B over
z sub p in r variables will be read from any host. p and the resultant
of A and B will be sent back to this host. ]

(1) [Initialize] BEGIN3; BEGINC; ISIZE := 72; RMARG := 72.
(2) [MPRES forever] repeat {
    t := IUNIT; IUNIT := ANY; BLOCK := TRUE;
      r := IREAD(); p := IREAD(); A* := LREAD(); B* := LREAD();
    IUNIT := t;

    C*:=MPRES(r,p,A*,B*);

    t := OUNIT; OUNIT := SRV(HOST);
      IWRITE(p); LWRITE(C*); WRITE; FIN;
    OUNIT := t } ||
```

Listing 1

Listing 2 shows the program, which uses this service to compute the resultant of two integral polynomials. The original algorithm IPRES reduces the problem of the computation of the resultant of a pair of integral polynomials to the problem of the computation of the resultant of several pairs of modular polynomials. Instead of calling MPRES, the distributed version of IPRES, XIPRES, writes the actual parameters for MPRES to the hosts, which offer MPRES as a service.

After the distribution of the work, XIPRES collects the results by reading from IUNIT = ANY in algorithm CLCTMR which uses the IPCRA (Integral Polynomial Chinese Remainder Algorithm) to construct the result of XIPRES.

```
safe BLOCK.

const ANY = 100.                                       [ unspecified host ]
const SRVCNT = 4.                                     [ Number of servers ]
safe  CURSRV.                          [ current unit number of server ]
const SRV(h) = 100 + h.                     [ host number to unit number ]
const HST(u) = u - 100.                     [ unit number to host number ]
const NXTSRV(u) = SRV(REM(HST(u),SRVCNT)+1).         [ next server number ]
```

```
                        MAIN.
[ Demonstrate a distributive version of IPRES ]

(1) [Initialize] BEGIN3; BEGINC; ISIZE:=72; RMARG:=72; CURSRV=SRV(1).
(2) [Random polynomials] CLOUT("#Vars(<=4) : ?"); WRITE; r := GREAD();
    CLOUT("'#Bits p degree_vec' for A : ?"); WRITE;
    k1 := GREAD(); q1 := RNREAD(); N1 := LREAD();
    CLOUT("'#Bits p degree_vec' for B : ?"); WRITE;
    k2 := GREAD(); q2 := RNREAD(); N2 := LREAD();
    A := IPRAN(r,k1,q1,N1); B := IPRAN(r,k2,q2,N2).
(3) [Resultant] C := XIPRES(r,A,B);
    r' := r-1; V := INV(REDUCT(("u","z","y","x"), 4-r'));
    IPWRIT(r',C,V); WRITE.
(4) END1 ||

                        C:=XIPRES(r,A,B)
[Integral polynomial resultant, distributed computation.
A and B are integral polynomials in r variables, r ge 1, of positive
degrees.  C is the resultant of A and B with respect to the main
variable, an integral polynomial in r-1 variables.]

     safe I,m,n,p.
(1)  [Compute coefficient bound.] d:=IPMAXN(r,A); e:=IPMAXN(r,B);
     m:=FIRST(A);  n:=FIRST(B);  d:=IEXP(d,n); e:=IEXP(e,m);
     f:=IFACTL(m+n);  f:=IPROD(e,f); f:=IPROD(d,f).
(2)  [Initialize.]  I:=PRIME; s := 0; Q := 1.
(3)  [Loop.] while I # () do { ADV(I;p,I); A* := MPHOM(r,p,A);
       if PDEG(A*)=m then { B* := MPHOM(r,p,B);
            if PDEG(B*)=n then {
                 t := OUNIT; OUNIT := CURSRV;
                   IWRITE(r); CWRITE(' '); IWRITE(p); WRITE;
                   LWRITE(A*); WRITE; LWRITE(B*); WRITE;
                 OUNIT := t;
                 CURSRV := NXTSRV(CURSRV); s := s+1;
                 Q := IPROD(Q,p);
                 if ICOMP(Q,f) >= 0 then { C := CLCTMR(s,r); return }
            } } } .
(4)  [Algorithm fails.]  CLOUT("ALGORITHM XIPRES FAILS.");
     EMPTOB; stop ||

                        C:=CLCTMR(s,r)
[Collect modular intermediate results for XIPRES
s is the number of data packets, which have been sent by XIPRES to other
hosts. r is the number of variables for the polynomials in XIPRES.
The s computed resultants are collected. ]

(1) [Init] Q := 1; C := 0; r' := r - 1.
(2) [Collect] t := IUNIT; IUNIT := ANY; BLOCK := TRUE;
    for i=1,...,s do { p := IREAD(); C* := LREAD();
      q := MDHOM(p,Q); q := MDINV(p,q); C := IPCRA(Q,p,q,r',C,C*);
      Q := IPROD(Q, p);
    }; IUNIT := t ||
```

Listing 2

Because the homomorphic images of the polynomials with MPHOM can also be computed independently, algorithm servers could have been introduced, which accept two integral polynomials and a list of prime numbers and send back a list of pairs, consisting of a prime number and the corresponding modular resultant, or which use the chinese remainder algorithm themselves to reply intermediate results needed by XIPRES to build the integral polynomial resultant. But these approaches would have not allowed the identification of an existing algorithm with a host.

Several kinds of communicating algorithms are possible to solve a given problem and the ultimate choice depends, amongst other criteria, on the amount of data which needs to be moved over the network in order to allow an algorithm server to start its computations.

7 Limitations

In addition to the usage of algebraic algorithms, the user has to deal with the administration of return host numbers associated with each set of data received from other hosts. In the case of one requestor, the host to which to return the data is held in the variable HOST. In order to send back intermediate results, which reached a host over several stations, the user has to keep track of the requestor host numbers.

Only in exceptional cases it not necessary do distinguish the computed intermediate results as in the example with MPRES. Several instances of the algorithm MPRES compute the intermediate results, which are used by the requestor in any order. In the general case, the user has to ensure that the intermediate results are available only at their intended places, because either there are several different algorithm servers involved in a computation or the order of the usage of intermediate results is important. In both cases reading with HOST=ANY might not be appropriate, and the user has to deal again with details not part of the algorithmic problem in a narrow sense.

Beside the fussy use of the objects and system calls offered by the operating system for interprocess communication, the implementation is not stable against run-time errors. The abnormal termination of an algorithms running on one host almost always corrupts the concurrent algorithm which is currently executing on the network. The user has to restart the algorithms on each host. But the user should not be bothered with the stability and the robustness of the system, which is used for the implementation of algebraic algorithms.

Further investigations have to be made in the development of tools for tracing the execution and for debugging of the various algorithms. In addition a tool for measuring times spent for execution and the transmission of data is not available.

8 Discussion

We have shown, how a user can write algorithms in the computer algebra system SAC-2, which use several hosts on a network for the execution of subalgorithms. Our approach treats hosts on a network as a special kind of peripheral devices, which can be accessed by setting the value for the input and output unit to the value associated with this machine.

The programmer has to partition the problem into nearly independent subproblems, describe the solution by communicating algorithms and install them on different machines on the network. The data to be transmitted over the network has not to be actual parameters for algorithms. Other kind of data, as the desired precision of intermediate results or status messages, can be sent as well.

Many details have to be considered which are not essential for the process of algorithm development, e.g. the specific placement of algorithms on, probably different sized, machines or the format of the transmitted data. In addition the user must take care about the sequence of parameters sent to and the results sent from other hosts, because random access to the content of the local buffer is not possible.

However, our concept is used to gain experience both in the development of parallel algorithms suitable for the execution on a network of loosely coupled machines and the development of language constructs which are suitable for the user of the SAC-2 system who wants to write such algorithms.

References

[1] Collins, G.E. (1971). The Calculation of Multivariate Polynomial Resultants, Journal of the ACM 19, 515–532.

[2] Fitch, J. (1988). Compiling for parallelism, First international workshop on Computer Algebra and Parallelism, Grenoble.

[3] Loos, R. (1988). SAC-2 Documentation, Universität Tübingen.

[4] Seitz, S. (1985). Ein Interpretierergenerator für Termersetzungssysteme, Diplomarbeit, Universität Karlsruhe.

[5] Sun Microsystems Inc. (1986), Networking on the Sun Workstation.

VLSI MANIPULATION OF POLYNOMIALS

Françoise SIEBERT-ROCH[1] and Jean-Michel MULLER[2]
CNRS - Laboratoire TIM3
Institut National Polytechnique de Grenoble
38031 GRENOBLE Cedex
FRANCE

ABSTRACT

This paper is devoted to the parallel and hardwired multiplication and Euclidean division of polynomials. After a survey of classical algorithms concerning that topic, we give a brief complexity study. Then we present some new algorithms and architectures suitable for VLSI implementation.

KEYWORDS

Computer Arithmetic, Computer Algebra, Special-purpose circuits, VLSI, Multiplication, Division, Polynomials.

---

[1]This work is partially supported by the "GRECO Calcul formel" of the french *Centre National de la Recherche Scientifique* (CNRS).

[2] and the "GRECO C^3" of the CNRS.

COMPUTER ALGEBRA AND PARALLELISM ISBN 0-12-221100-6

I - INTRODUCTION.

The huge execution times obtained in computer algebra are essentially due to the basic operations slowness. The calculus of multivariate polynomials GCDs, for example, leads to an intermediate coefficient swell which can give some great execution time. Now, the GCD is an indispensable tool to work in $\mathbb{Q}[X_1, \ldots, X_n]$. Thus, an improvement of the basic exact arithmetic operations would decrease consequently the cost of most of computer algebra algorithms. In the current systems, infinite precision and polynomial arithmetic are performed in software, which is the main cause for slowness. In this paper, we propose some new architectures for computer algebra to do the basic operations at the hardware level.

Since the integration abilities have increased drastically with current VLSI technologies, more and more complex algorithms can be integrated nowadays. This progression can be given by some examples :

The only arithmetic operation implemented in the first microprocessors was addition (for instance, the instruction set of the Zilog Z80 did not contain multiplication : this operation had to be done by software). In the following generation of microprocessors, multiplication was given in the instruction set, but not division. Nowadays, the four basic arithmetic operations are integrated : circuits designers try to implement more complicated functions like Sorting and Fast Fourier Transform [25] and elementary functions [8], [17], [19], [27]. It seems natural nowadays to design algorithms in order to implement infinite-precision arithmetic and polynomial manipulation.

Polynomial multiplication and division constitute the aim of this paper. In part II we recall classical algorithms used to multiply two polynomials, part III is devoted to a complexity study, in part IV and V we present new hardware algorithms and special purposed architectures for multiplying and dividing polynomials.

II - CLASSICAL ALGORITHMS FOR POLYNOMIAL MULTIPLICATION.

The performances of multiplication algorithms are linked to the representation and whether the polynomials are sparse or not, thus the methods presented below are more or less interesting according as the polynomials are dense or sparse.

Let $P = \sum_{i=0}^{n} p_i X^{e_i}$ and $Q = \sum_{j=0}^{m} q_j X^{f_j}$

The naive method in order to compute PQ would consist in generating all the mn terms $p_i q_j X^{e_i+f_j}$, sorting them by degree order (in the case of sparse polynomials) and then treating the terms of equal degrees. The complexity of such a method is $\Theta(mn \log(mn))$.
We may consider several approach to improve this algorithm :

- The first class of algorithms attempts to minimize the number of exponent comparisons to be performed. These methods are particularly adapted to sparse polynomials whose product exponents are *a priori* unknown.

If we consider the product as a sum of n+1 polynomials (if $n \leq m$):

$$(\sum_{i=0}^{n} p_i X^{e_i}) * (\sum_{j=0}^{m} q_j X^{f_j}) = \sum_{i=0}^{n} (\sum_{j=0}^{m} p_i q_j X^{e_i + f_j})$$

then the calculus of each term of the sum needs no sorting neither merging.
It can be shown (see [10]) that the cost of this method is of $\Theta(n^2 m)$, or $\Theta(mn \log(n))$ using a "divide and conquer" sum.

Some algorithms attempt to generate the result step by step, by decreasing order of the degree. They need sorting algorithms to find the next couple of exponent to be considered. The heapsort algorithm [10] used in the ALTRAN system, reduces the number of comparisons to $\Theta(mn \log(n))$. In the ALTRAN system the polynomial representation is sparse.

● Other techniques try to minimize the number of coefficient multiplications. The Algorithm of Karatsuba (1962) computes the product of two polynomials of degrees approximately 2^k:

$$P = X^{2^{k-1}} P_1 + P_2 \quad \text{and} \quad Q = X^{2^{k-1}} Q_1 + Q_2$$

It is based upon the above relations :

$$P\,Q = X^{2^k} (P_1\,Q_1) + X^{2^{k-1}} (P_1\,Q_2 + P_2\,Q_1) + P_2\,Q_2$$

$$= X^{2^k} (P_1\,Q_1) + X^{2^{k-1}} ((P_1 + P_2)\,(Q_1 + Q_2) - P_1\,Q_1 - P_2\,Q_2) + P_2\,Q_2$$

In the second relation the terms $P_1\,Q_1$ and $P_2\,Q_2$ appear twice and the multiplication requires only three multiplications. Applying this scheme recursively, the multiplication of two polynomials of degree n requires $n^{\log_2 3}$ multiplications.

● The last class of algorithms uses evaluation of the product on a certain number of points and then computation of the product by interpolation.

FFT algorithm [1] uses evaluation at points ω^i, $i \le n$, where ω is a $(n+1)^{st}$ root of unity different from 1, and n is the sum of the degrees of P and Q.

Let us suppose $P = \sum_{i=0}^{k} p_i X^i$ and $Q = \sum_{j=0}^{l} q_j X^j$ (this method is convenient for *dense* polynomials). Let n = k+l, and let us call Ω the (n+1)x(n+1) matrix with coefficient $\omega^{(i-1)(j-1)}$ at line i and column j. If we compute the matrix-vector product $\Omega \times (p_0\ p_1\ \dots\ p_k\ 0\ 0\ \dots\ 0)^t$, we obtain $(b_0\ b_1\ \dots\ b_n)^t$ where b_i is the value of P at point ω^i : we have evaluated P. The vector $(b_0\ b_1\ \dots\ b_n)^t$ is the *discrete fourier transform* of order n of the vector $(p_0\ p_1\ \dots\ p_k\ 0\ 0\ \dots\ 0)^t$.

Of course, practically, the coefficients b_i are computed using the Fast Fourier Transform (FFT) algorithm of Runge and Konitz ([20], 1924), rediscovered by Cooley and Tuckey ([7], 1965), which is faster than the matrix-vector product. Since the Fourier transform *evaluates* a polynomial at n points, the *inverse* Fourier transform (which may be computed using the same FFT algorithm) *interpolates* n values of a function at points ω^i by a polynomial. It is the basic principle of the following algorithm. For convenience, we shall note $p_i = 0$ for $i > k$, $q_j = 0$ for $j > l$.

Algorithm Product–FFT.

Step 1 (evaluation).
Compute $(b_0\ b_1\ \dots\ b_n)^t$ the Fourier transform of $(p_0\ p_1\ \dots\ p_n)^t$ and $(c_0\ c_1\ \dots\ c_n)^t$ the Fourier transform of $(q_0\ q_1\ \dots\ q_n)^t$ using the FFT algorithm.

Step 2 (products)
Compute the terms $r_i = b_i c_i$, $i = 0 \dots n$.

Step 3 (interpolation).
Compute the vector $(z_0\ z_1\ \dots\ z_n)^t$ which is the inverse Fourier transform of the vector $(r_0\ r_1\ \dots\ r_n)^t$. z_i is the coefficient of degree i of PQ.

The complexity of this method is that of FFT algorithm : $\Theta(n \log n)$.

III - COMPLEXITY STUDY.

In circuitry design, since the chip area is a good measure of cost, conception of VLSI circuits needs a trade-off between computation time and area. Time complexity of the most usual arithmetic operation by circuits has been known for a long time. In [28] Winograd introduced the concept of (d,r) elements, and used it to find time complexity of addition [28] and multiplication [29]. More recently, some authors found "mixed-complexity" results of the form AT^{α} (A = area, T = time) for a lot of problems (see [2] for multiplication, [25] for FFT and sorting). In [26], Vuillemin gave such a result for a class of problems including polynomial multiplication. Here, we shall give a result of complexity for a *parallel* machine (which is not necessary a circuit, we can use a 3-dimensional machine), between the time of computation and the number of processors used. We start this study from Winograd's results.

In this study it is assumed that the functions are constructed with circuit comprising only denary circuit elements with r inputs at most. These elements are called (d,r) elements ([24], [28], [29]). We shall suppose that these elements work in the same delay, and we shall choose this delay as time unit. A circuit may be considered as a graph, which vertices are (d,r) elements, and which edges are wires connections. We shall suppose that there is no cycle in the circuit's graph.

Let E be a set, let F be a function from E^n to E. F will be said *absolutely dependent* if for any $i \in \{1,2, \ldots, n\}$ there exists $(x_1, \ldots, x_n) \in E^n$ and $x^* \neq x_i$ such that :

$$F(x_1, \ldots, x_{i-1}, x_i, x_{i+1}, \ldots, x_n) \neq F(x_1, \ldots, x_{i-1}, x^*, x_{i+1}, \ldots, x_n).$$

An absolutely dependent function is a function whose computation by a (d,r) circuits needs the examination of *all* entries.

Example : $F : \mathbb{R}^{2n} \rightarrow \mathbb{R}$

$(a_0, a_1, \ldots, a_{n-1}, b_0, b_1, \ldots, b_{n-1}) \rightarrow m_{n-1}$

Where m_{n-1} is the coefficient of degree n-1 of the product of polynomials :

$$(a_0 + a_1X + \ldots + a_{n-1}X^{n-1})\ (b_0 + b_1X + \ldots + b_{n-1}X^{n-1}).$$

(m_{n-1} is equal to $\sum_{k=0}^{n-1} a_k b_{n-1-k}$)

We shall suppose that in our circuits, the data flow is independent from the data values. Such a data flow may be represented by a (d,r) circuit, which input vertices correspond to the entries of the function computed, and which output vertice corresponds to the value of the function computed. (Fig.1).

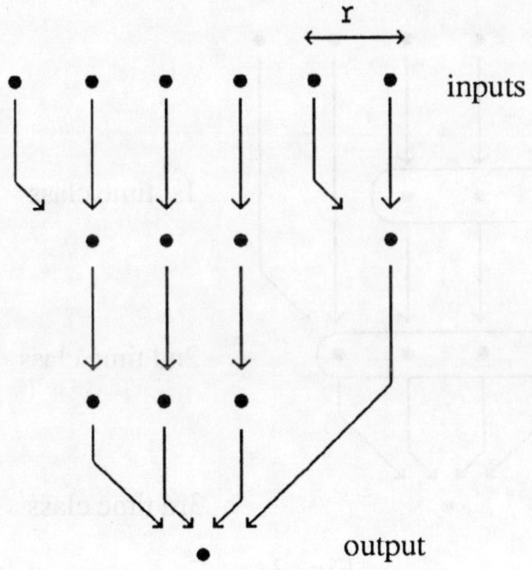

Fig. 1

A data flow represented by a (d,r) circuit may be mapped in several ways. For instance, if we call ith time class the class of (d,r) elements which work at time i, Fig. 2 and 3 represent two different possible mappings of the (d,r) circuit of Fig. 1.

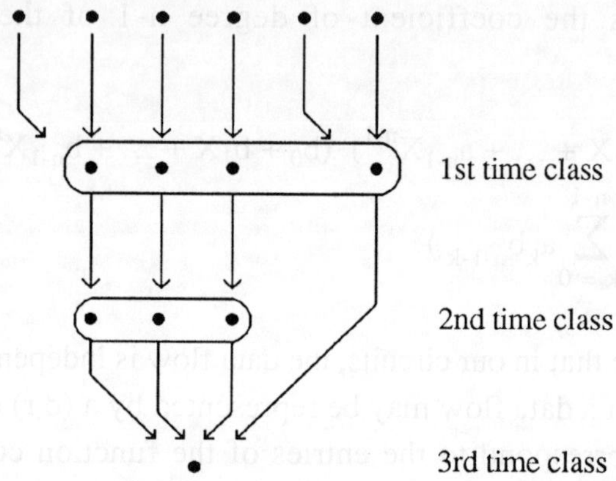

Fig.2

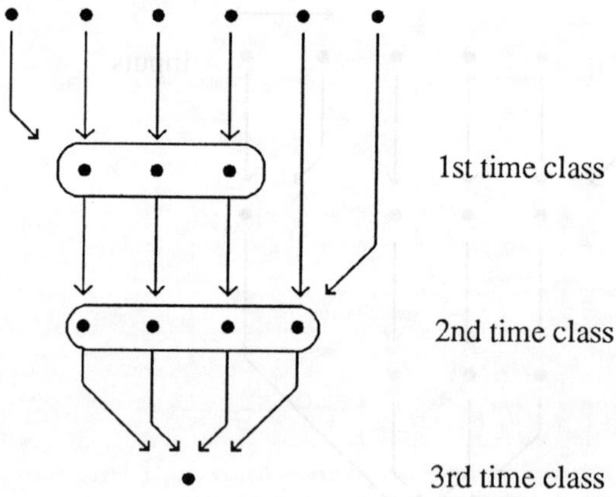

Fig. 3

Now let us assume that we implement an algorithm with a machine containing p processors which can work in parallel. A mapping of a (d,r) graph can be implemented if and only if all its time classes have less than p vertices. We shall call *width* of a mapping the number of vertices of it's largest i^{th} time class, and *depth* of this mapping the number of its time classes . Let us consider the following problem : we have to compute an absolutely dependent function of n entries with that machine. The best evaluation time of this function is the minimal depth among the dephts of the mappings of width lower than or equal to p. Let us consider the output node of a (d,r) graph associated to the computation of an absolutely dependent function. We shall consider that this node ends its calculation at time t = 0. Since this node has at most r inputs (it is a (d,r) element); the -1st time class contains at most r elements. If the -k^{th} time class contains m elements then :

- if $m \leq p$ then each element may be a node, thus there is at most rm elements (the inputs of this node) in the -$(k+1)^{st}$ class.
- if $m > p$, then p elements may be a node : there is at most pr + m - p elements in the -$(k+1)^{st}$ class.

Thus it is obvious to prove by induction than the number of elements of the -k^{th}time class is bounded by u_k where

$$u_0 = 1 \qquad u_{k+1} = \begin{cases} r\,u_k & \text{if } u_k \leq p \\ p\,(r-1) + u_k & \text{if } u_k \geq p \end{cases}$$

Since the function of n variables which is computed is absolutely dependent, all its entries must be examinated by (d, r) elements. Thus the time of computation of such a function is at least k, where $k = \min \{ k / u_k \geq n \}$ (let us recall that the time unit is the delay of a (d, r) element). Therefore :

- if $n \leq r\,p$ then $k = \lceil \log_r (n) \rceil$
- else $k = \left\lfloor \frac{n}{p\,(r-1)} \right\rfloor + \left\lceil \log_r \left(n - \left\lfloor \frac{n}{p\,(r-1)} \right\rfloor - p\,(r-1)\right) \right\rceil$

If we compute a product $(a_0 + \ldots + a_{n-1}X^{n-1}) \times (b_0 + \ldots + b_{m-1}X^{m-1})$, with $m \leq n$, the coefficient of degree n of the product is an absolutely dependent function of $(a_0, a_1, a_2, \ldots, a_{n-1}, b_0, b_1, b_2, \ldots, b_{m-1})$. Thus we can apply the preceding result, and we obtain :

THEOREM. A parallel computer with p processors which are assimilable to (d,r) elements cannot compute the product of two polynomials of degrees m and n ($n \geq m$) in a time better than :

- if $n \leq r p$: $\lceil \log_r (n) \rceil \tau$
- else : $\lfloor \frac{n}{p(r-1)} \rfloor + \lceil \log_r (n - \lfloor \frac{n}{p(r-1)} - p(r-1) \rfloor) \rceil \tau$

where τ is the delay of one processor.

IV - HARDWIRED MULTIPLICATION OF POLYNOMIALS.

Let us assume that we want the product $P = p_{2n}X^{2n} + \ldots + p_0$ of two dense polynomials :

$$A = a_nX^n + \ldots + a_0 \qquad B = b_nX^n + \ldots + b_0$$

Since for large polynomials it is irrelevant to suppose that all the coefficients a_i and b_j, $0 \leq i,j \leq n$, enter the circuit in parallel, we shall suppose that these coefficients come in a *serial fashion* : at time $T = i$, a_i and b_i become available (the coefficients are available lower degrees first, a solution similar to that presented here may be built for serial polynomial multiplier with coefficients available higher degrees first). Our purpose is to start the computation before having all the coefficients. Our algorithm is derived from an algorithm designed by Chen and Willoner ([5]) in order to multiply radix-2 integers. Let us point out :

$$\begin{cases} A^{(0)} = a_0 \\ B^{(0)} = b_0 \\ A^{(i)} = a_i X^i + a_{i-1} X^{i-1} + \ldots + a_0 = a_i X^i + A^{(i-1)} \\ B^{(i)} = b_i X^i + b_{i-1} X^{i-1} + \ldots + b_0 = b_i X^i + B^{(i-1)} \\ P^{(i)} = A^{(i)} B^{(i)} \end{cases}$$

Starting from $P^{(0)} = a_0 b_0$, we shall compute recursively $P^{(1)}$, $P^{(2)}$, $P^{(n)}$ = AB, using the following recursion formula :

$$P^{(i)} = A^{(i)} B^{(i)} = P^{(i-1)} + b_i A^{(i-1)} X^i + a_i B^{(i)} X^i.$$

This recursion is implemented in the circuit of Fig. 4. Fig. 5 illustrates the behavior of that circuit. The terms of the product appear in the *addition cells* (see Fig. 4). This circuit is built with some different kinds of cells :

1– Multiplier cells (on line multipliers – see [9], [5], [13] and [14] – seem to be a good compromise between time of computation and size). Moreover, such multipliers involve a simplified layout.

2– Memory cells.

3– Addition cells.

C and New are two logical values which are set to 1 at the beginning of a computation, and to zero immediately after.

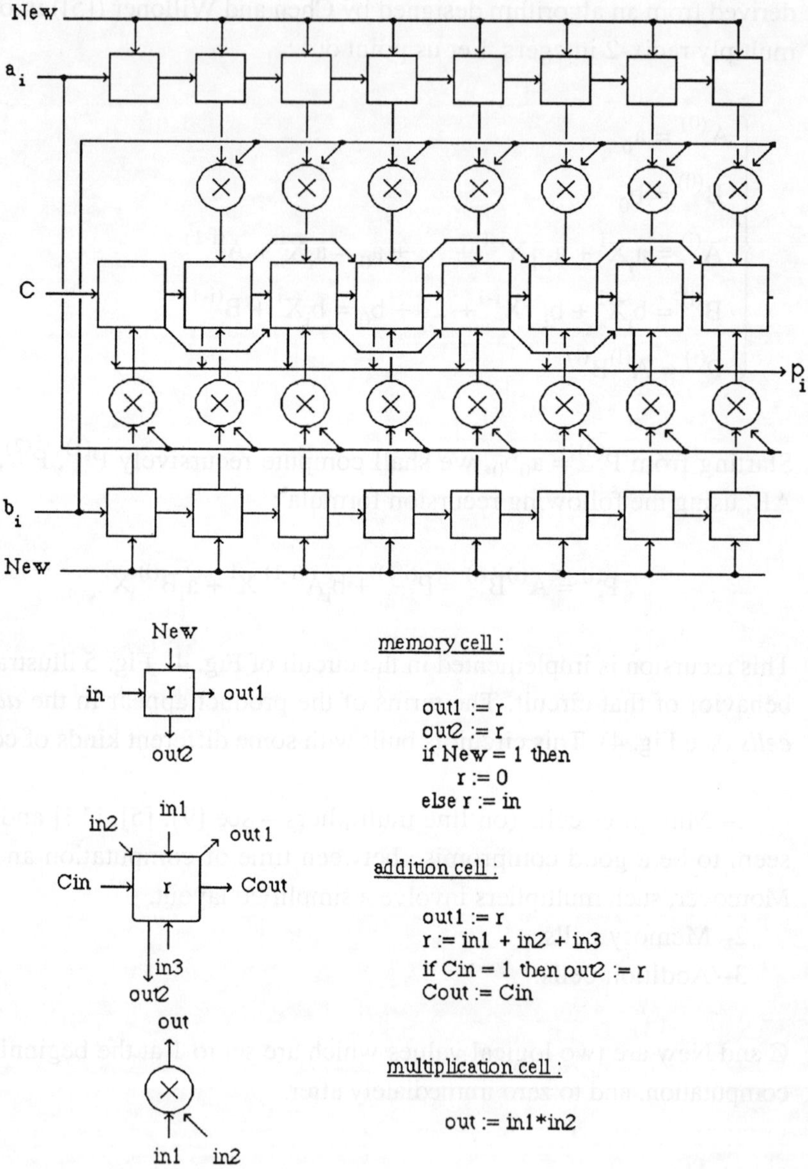
New
a_i
C
p_i
b_i
New
New
in
r
out1
out2
memory cell :
out1 := r
out2 := r
if New = 1 then
r := 0
else r := in
in1
in2
out1
Cin
r
Cout
in3
out2
addition cell :
out1 := r
r := in1 + in2 + in3
if Cin = 1 then out2 := r
Cout := Cin
out
in1
in2
multiplication cell :
out := in1*in2

Fig. 4

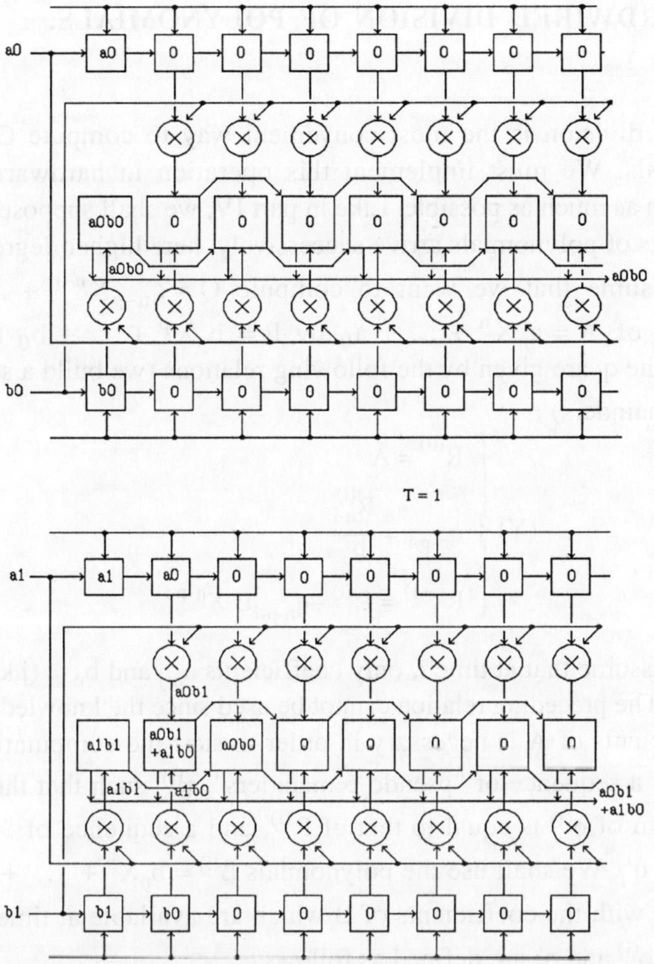

Fig. 5

A symmetry argument can easily show that if we bring to the multiplier the coefficients of the factors in a reverse order, "high degrees first", the circuit computes the correct product, and outputs it in the same reverse order. Thus our "low degrees first" multiplier is also a "high degrees first" multiplier. It is important, because, since the divider presented in the last part of this paper is a "high degrees first" divider, one may prefer to chose a data flow "from the highest degree coefficient to the lowest one".

V - HARDWIRED DIVISION OF POLYNOMIALS.

Euclidean division is the most convenient way to compute GCDs of polynomials. We must implement this operation in hardware, using parallelism as much as possible. Like in part IV, we shall suppose that the coefficients of polynomials arrive successively, here higher degrees first. Let us assume that we want to compute $Q = q_{n-p}X^{n-p} + \dots + q_0$, quotient of $A = a_nX^n + \dots + a_0$ by $B = b_pX^p + \dots + b_0$ $(p \leq n)$. Usually, the q_i are given by the following relations (we build a sequence $R^{(i)}$ of remainders) :

$$(Y)\begin{cases} R^{(0)} = A \\ q_{n-p-i} = \dfrac{R^{(i)}_{n-i}}{b_p} \\ R^{(i+1)} = R^{(i)} - q_{n-p-i}B\,X^{n-p-i} \end{cases}$$

We shall assume that at time i, only coefficients a_{n-j} and b_{p-k} $(j,k \leq i)$ are available. The preceding relation cannot be used since the knowledge of all the coefficients of A is necessary in order to start the computation : we shall build a sequence of "pseudo-remainders", $\phi^{(i)}$, such that the higher degree term of $\phi^{(i)}$ is equal to that of $R^{(i)}$, and a sequence of "pseudo-quotients" q'_i. We shall use the polynomials $\beta^{(i)} = b_pX^p + \dots + b_{p-i}X^{p-i}$ (β_i is built with the coefficients of B which are available at time i). The sequences $\phi^{(i)}$ and q'_i are defined as follows :

$$(Y')\begin{cases} \phi^{(0)} = a_nX^n \\ q'_{n-p-i} = \dfrac{\phi^{(i)}_{n-i}}{b_p} \\ \phi^{(i+1)} = \phi^{(i)} - q'_{n-p-i}\,\beta^{(i+1)}X^{n-p-i} + a_{n-i-1}X^{n-i-1} - \sum_{k=0}^{i-1} q'_{n-p-k}b_{p-i-1}X^{n-i-k-1} \end{cases}$$

Of course, the coefficients of each degree may be computed in parallel. The following theorem proves that (Y) and (Y') algorithms give the same quotient.

Theorem.

For any j, $0 \leq j \leq n-p$, $q_{n-p-j} = q'_{n-p-j}$.

Proof.

We shall prove this result by induction. It is obviously true for $j = 0$. Let us assume that it is true for any $j \leq i$, i given. We shall use the following symbols :

$$\delta_{i,j} = \begin{cases} 1 \text{ if } i = j \\ 0 \text{ if } i \neq j \end{cases} \qquad \Delta_{i,j} = \begin{cases} 1 \text{ if } i \geq j \\ 0 \text{ if } i < j \end{cases}$$

Since $R^{(i+1)}$ is equal to $R^{(i)} - q_{n-p-i}BX^{n-p-i}$, the coefficient of degree j of $R^{(i+1)}$, $R_j^{(i+1)}$ is equal to $R_j^{(i)} - q_{n-p-i}b_{j-n+p+i}$. Thus we obtain :

$$R_j^{(i+1)} = a_j - \sum_{k=0}^{i} q_{n-p-k}b_{j-n+p+k}$$

We deduce from that result the value of the coefficient of highest degree of $R^{(i+1)}$:

$$R_{n-i-1}^{(i+1)} = a_{n-i-1} - \sum_{k=0}^{i} q_{n-p-k}b_{p+k-i-1}$$

Now, let us compute the coefficient of highest degree of $\phi^{(i+1)}$. We have :

$$\phi^{(i+1)} = \phi^{(i)} - q'_{n-p-i}\,\beta^{(i+1)}X^{n-p-i} + a_{n-i-1}X^{n-i-1} - \sum_{k=0}^{i-1} q'_{n-p-k}b_{p-i-1}X^{n-i-k-1}$$

thus the coefficient of degree j of $\phi^{(i+1)}$, $\phi_j^{(i+1)}$ is equal to

$$\phi_j^{(i)} - q'_{n-p-i}b_{j-n+p+i}\,\Delta_{j-n+p+i,p-i-1}\Delta_{j,n-p-i} + a_{n-i-1}\delta_{n-i-1,j} - q'_{i+j+1-p}b_{p-i-1}\,\Delta_{i+j+1-p,n-p-i+1}$$

(recall that at time i+1, only b_p, b_{p-1}, b_{p-i-1} are available, and that the q_k, $k \leq n-p-i$ are not yet computed).

$$= \phi_j^{(0)} - \sum_{k=0}^{i} q'_{n-p-k} b_{j-n+p-k} \Delta_{j-n+k,-k-1} \Delta_{j,n-p-k} + \sum_{k=0}^{i} a_{n-k-1} \delta_{n-k-1,j} - \sum_{k=0}^{i} q'_{k+j+1-p} b_{p-k-1} \Delta_{2k+j,n}$$

thus, the coefficient of highest degree of $\phi^{(i+1)}$ is equal to :

$$\phi_{n-i-1}^{(i+1)} = - \sum_{k=0}^{i} q'_{n-p-k} b_{p+k-i-1} \Delta_{k-i,-k} \Delta_{n-i-1,n-p-k} + a_{n-i-1} - \sum_{k=0}^{i} q'_{k+n-i-p} b_{p-k-1} \Delta_{2k,i+1}$$

$$= a_{n-i-1} - \left[\sum_{\substack{k=0 \\ 2k \geq i}}^{i} q'_{n-p-k} b_{p+k-i-1} + \sum_{\substack{k=0 \\ 2k \leq i-1}}^{i} q'_{n-p-k} b_{p+k-i-1} \right]$$

$$= a_{n-i-1} - \sum_{k=0}^{i} q'_{n-p-k}\, b_{p+k-i-1}$$

Thus, since $q'_{n-p-k} = q_{n-p-k}$ for each $k \leq i$ (it is the induction hypothesis), we deduce that $R_{n-i-1}^{(i+1)} = \phi_{n-i-1}^{(i+1)}$ thus :

$$q_{n-p-i-1} = \frac{R_{n-i-1}^{(i+1)}}{b_p} = -\frac{\phi_{n-i-1}^{(i+1)}}{b_p} = q'_{n-p-i-1}$$

And the theorem is proved. ■

The circuit presented Fig.6 implements recursion (Y').

Now, in order to explain the behavior of that circuit, let us consider an example : we compute the quotient of $A = 4X^5 - 3X^4 + 0\,X^3 + 2X^2 + 6X - 4$ by $B = X^3 + 2X^2 - 3X + 1$. This quotient is equal to $4X^2 - 11X + 34$. Recursion (Y) gives the following remainders :

$$R^{(1)} = -11\,X^4 + 12\,X^3 - 2X^2 + 6X - 4$$
$$R^{(2)} = 34\,X^3 - 35\,X^2 + 17\,X - 4$$
$$R^{(3)} = -103\,X^2 + 119\,X - 38$$

And recursion (Y') gives the following pseudo-remainders :

$$\phi^{(0)} = 4\,X^5$$
$$\phi^{(1)} = -11\,X^4$$
$$\phi^{(2)} = 34X^3 - 33\,X^2$$
$$\phi^{(3)} = -103X^2 + 113\,X - 34$$

The circuit of Fig. 6 contains different kinds of cells :

- Multiplication cells and a division cell (serious advantages in circuit layout, circuit area and computation time may be obtained using on-line multipliers and divider, like those described in [9], which are very small, simple and efficient, and lead to a simple layout). We suppose that these cells have the same delay.
- Register cells : "b cells" and "q cells", which contains the successive coefficients of B and Q.
- "ϕ cells", which compute the successive values of the pseudo remainders $\phi^{(i)}$.
- Delay cells, needed for synchronization, with the same delay as multiplication and division cells.
- C is a "starting signal" which is set to 1 at the beginning of a computation, and to 0 immediately after. Fig. 7 presents the inputs and outputs of the circuit during the first 3 steps of calculation. Fig. 8 presents the different data values in the circuit during the computation for the example presented above (quotient of $A = 4X^5 - 3X^4 + 0\,X^3 + 2X^2 + 6X - 4$ by $B = X^3 + 2X^2 - 3X + 1$).

This circuit computes the quotient of a polynomial A of degree n by a polynomial B of degree p in time Θ (n-p), since at each step of (Y') a new coefficient of that quotient is obtained. The "true remainder" of the Euclidean division is obtain by adding the last "pseudo remainder" $\phi^{(n-p+1)}$ to the polynomial built with the unused coefficients of A : $a_{n-p-1}X^{n-p-1} + a_{n-p-2}X^{n-p-2} + \ldots + a_0$.

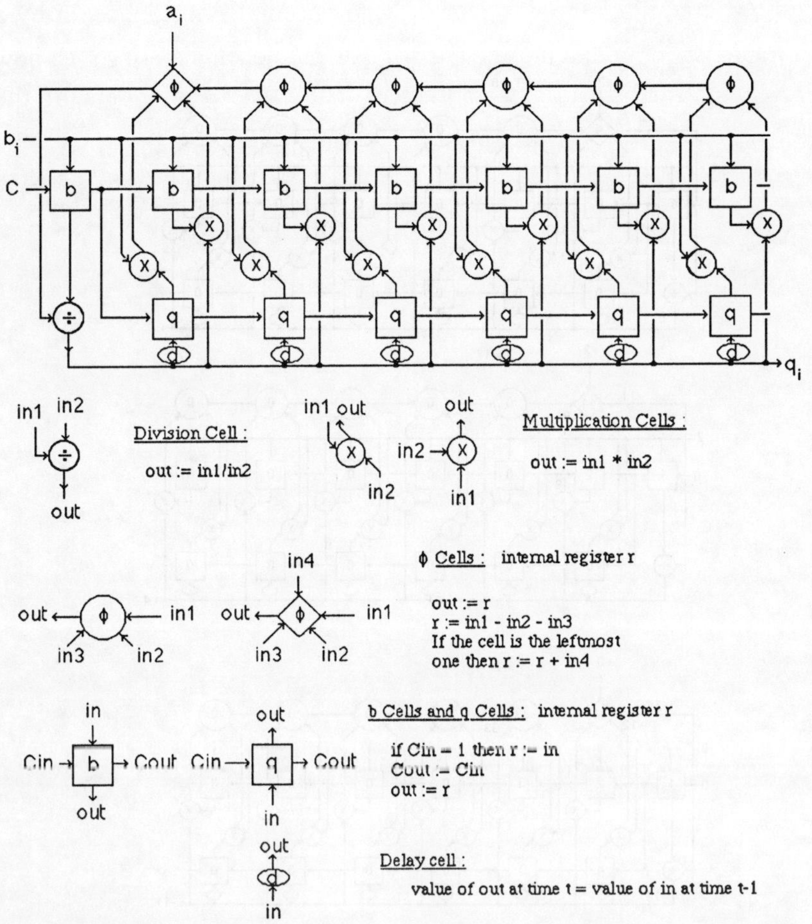

a_i
b_i
C
q_i
in1 in2
out
Division Cell :
out := in1/in2
in1 out
in2
out
in2
in1
Multiplication Cells :
out := in1 * in2
in4
out in1 in3 in2
out in1 in3 in2
ϕ Cells : internal register r
out := r
r := in1 - in2 - in3
If the cell is the leftmost
one then r := r + in4
in
Cin b Cout
out
out
Cin q Cout
in
b Cells and q Cells : internal register r
if Cin = 1 then r := in
Cout := Cin
out := r
out
in
Delay cell :
value of out at time t = value of in at time t-1

Fig. 6

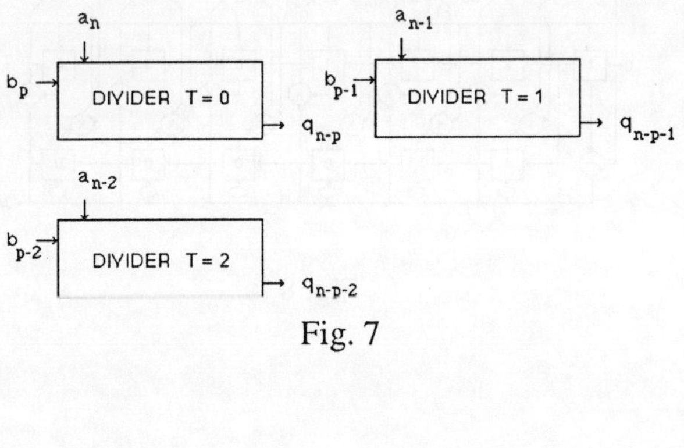

a_n
b_p
DIVIDER T = 0
q_{n-p}
a_{n-1}
b_{p-1}
DIVIDER T = 1
q_{n-p-1}
a_{n-2}
b_{p-2}
DIVIDER T = 2
q_{n-p-2}

Fig. 7

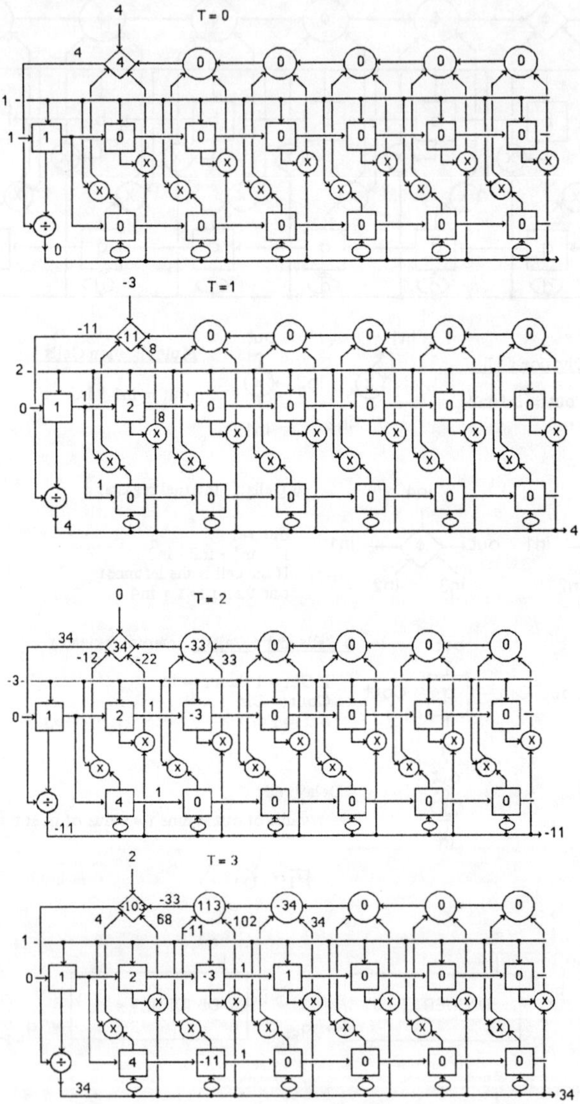

Fig. 8

V - PERFORMANCES.

lier cells and the divider cell of our circuits are on-line architectures ([9], [14], [21]). Our polynomial divider and multiplier become on-line circuits. A good measure of the performances of on-line circuits is there *delay* (an on-line algorithm is said to have a delay δ if for the generation of the j-th digit of the result, $j+\delta-1$ digits of the corresponding operands are required). In [9], Ercegovac and Trivedi give a division algorithm with an on-line delay $\delta = 4$ (which may be perhaps improved). It is well known that the other arithmetic operations (addition, subtraction, multiplication and square root) have an on-line delay $\delta = 1$. If we assume that our divider cell uses Ercegovac and Trivedi's algorithm, we can easily evaluate the delay of our architectures.

Polynomial multiplier. The computation of the first coefficient of a product of polynomials needs only a multiplication. Thus the on-line delay of our polynomial multiplier is 1 (the same as that of an integer multiplier).

Polynomial divider. The computation of the first coefficient of a quotient of polynomials needs an addition (in a Φ cell) and a division. Thus the on-line delay of our polynomial divider is 5.

REFERENCES.

[1] A.V. Aho, J.E. Hopcroft and J.D. Ullman (1974), The design and analysis of computer algorithms, Addison-Wesley publishing company.

[2] R.P. Brent et H.T. Kung (1981), The area-time complexity of binary multiplication, Journ. of the ACM, Vol.28 No3, pp. 521-534.

[3] A. Burks, H.H. Goldstine and J. Von Neumann (1946), Preliminary discussion of the logic design of an electronic computing instrument , Institute for advanced study, Princeton, N.J.; reprinted by C.G. Bell : Computer structures : readings and examples , Mc Graw-Hill, New-York, 1971.

[4] J.J.F. Cavanagh (1984), Digital computer arithmetic, design and implementation , Mc Graw-Hill Computer Science Series.

[5] I.N. Chen and R. Willoner (1979), An O(n) parallel multiplier with bit-sequential input and output, IEEE Trans. on Computers, Vol. C-28, No 10.

[6] G.E. Collins, (1966), PM, A system for polynomial manipulation, Comm. of the ACM, Vol. 9, No 8.

[7] J.W. Cooley and J.W. Tuckey (1965), An algorithm for the machine calculation of complex Fourier Series, Math. of Computation, Vol. 19, pp. 297-301.

[8] M.D. Ercegovac (June 1973), Radix 16 evaluation of certain elementary fonctions, IEEE Transactions on Computers.

[9] M.D. Ercegovac and K.S. Trivedi (1977), On line algorithms for division and multiplication, IEEE Transactions on Computers, Vol. C-26, pp. 681-687.

[10] W.M. Gentleman and S.C. Johnson (1976), Analysis of algorithms, a case study : determinants of matrices with polynomial entries, ACM Trans. on Math. Software, Vol. 2, No 3, pp 232-241.

[11] R. Gnanasekaran (1983), On a bit-serial input and bit-serial output multiplier , IEEE Trans. on Computers, Vol. C-32.

[12] A.L. Grnarov and M.D. Ercegovac, On the performance of on-line arithmetic, Proc. 1980 Intern. Conference on parallel processing, IEEE Publ. No 80CH1569-3, pp. 55-62.

[13] K. Hwang (1979), Computer arithmetic principles, architecture and design , New-York, J. Wiley&Sons Inc.

[14] M.J. Irwin (1978), A pipelined processing unit for on-line division, Proc. 5th on Computer architecture, IEEE Publ. No 78CH1284-9C, pp. 24-30.

[15] M. Kaminski (1986), An algorithm for polynomial multiplication which does not depend on the ring constants, Tech. report #418, Technion - Israel institute of technology, Computer Sci. Dep.

[16] D.E. Knuth (1969), The art of computer programming , Vol.2, Semmi-numerical algorithms, Addison Wesley, Reading, Mass.

[17] B.G. De Lugish (1970), A class of algorithms for automatic evaluation of certain elementary functions in a binary computer, Ph. D. Dissertation, Depart. of Computer Sci., Univ. of Illinois.

[18] C. Mead et L. Conway (1980), Introduction to VLSI systems, Ed. Addison-Wesley.

[19] J.M. Muller (1985), Discrete basis and computation of elementary functions, IEEE Trans. on computers, Vol. C-34 N° 9.

[20] C. Runge and H. Konitz (1924), Vorlesungen über numerisces reschen, Springer Verlag.

[21] T. Rhyne and N.R. Strader (1986), A signed bit-sequential multiplier, IEEE Trans. on computers, Vol. C-35 N° 10.

[22] N.R. Scott (1985), Computer systems and arithmetic, Prentice-Hall Inc., Englewood Cliffs, New Jersey.

[23] H.J. Sips (1984), Bit-sequential arithmetic for parallel processors, IEEE Transactions on Computers, Vol. C-33 N° 1.

[24] P.M. Spira (1973), Computation times of arithmetic and boolean functions in (d,r) circuits, IEEE Trans. on Computers, Vol. C-22 No 6.

[25] C.D. Thompson (1980), A complexity theory for VLSI, Ph.D. dissertation, Carneggie-Mellon University, Pittsburg, Pennsylvania.

[26] J.E. Vuillemin (1983), A Combinatorial limit to the computing power of VLSI circuits, IEEE Trans. on Computers, Vol. C-32 N° 3.

[27] J. Walther (1971), A unified algorithm for elementary functions, Joint Computer conference proceedings, Vol.38.

[28] S. Winograd (1965), On the time required to perform addition, Journ. of the ACM, Vol. 12, pp. 277-285.

[29] S. Winograd (1967), On the time required to perform multiplication, Journ. of the ACM, Vol. 14 N° 4, pp. 793-802.